[THIRTY-FIFTH THOUSAND.]

Intemperance :

ITS BEARING UPON AGRICULTURE ;

WITH AN

APPENDIX

CONTAINING THE TESTIMONY OF

LANDLORDS, FARMERS, LABOURERS, TRAVELLERS,
SCIENCE,

WITH AN APPEAL TO THE CLERGY, &c.,

BY

JOHN ABBEY.

" There is not, at the present day, any question, in my opinion, which
so deeply touches the Moral, the Physical, and the Religious welfare of the
World as the question of Temperance."—LORD CAIRNS.

THIRD EDITION REVISED.

PRICE SIXPENCE.

LONDON:
NATIONAL TEMPERANCE PUBLICATION DEPÔT, 337, STRAND. W.C.
EDINBURGH:
JOHN MENZIES & Co., 12, HANOVER STREET.
BELFAST:
IRISH TEMPERANCE LEAGUE. 18, LOMBARD STREET.

PREFACE.

My reason for issuing this Pamphlet is that for a long time I have desired to see this subject brought before the public, and I had hoped that some person far better qualified to deal with it than myself would have taken it up, but no one has done so at present (so far as I know). The major part of the following pages was intended to have been given in a speech, which I was asked to make at a Conference held in the Town Hall, Macclesfield, on Tuesday, June 14th, 1881, under the presidency of the Right Rev. Bishop Kelly, the subject for discussion being "Intemperance: its bearing upon Agriculture; with special reference to Beer in the Hay and Harvest Field;" but, in consequence of the number of speakers, the time allowed was only sufficient to enable me to say a few words on the subject. I have, therefore, on the advice of several friends deeply interested, thought it well to print it, with the valuable testimony contained in the Appendix, which I now humbly yet very respectfully dedicate with all its faults and failings to the Bishops and Clergy of the Church of England, and to the Agriculturists of the United Kingdom of Great Britain and Ireland.

JOHN ABBEY.

44, St. Giles', Oxford,
July, 1881.

THIRD EDITION.

Valuable testimony has been added to this Edition. I earnestly commend the subject to the careful study of the Clergy, Landlords, Farmers and other Employers.

June, 1882. J. A.

HOW THE CLERGY MAY HELP.

The Rev. J. T. Penrose, Rector of Gawsworth, said, "It was his intention to invite the farmers in his own parish to a conference on the subject, and see if some joint course of action could not be agreed upon. Unless something of this kind was done in the different parishes, he thought the movement could not meet with that success it otherwise would. If the farmers and labourers could meet together in conference he was sure much good would be the result.—*Speech at Macclesfield Conference.*

Intemperance :

ITS BEARING UPON AGRICULTURE.

———•••———

THE question of Intemperance and its bearing upon Agriculture is a subject of great and increasing importance, and one that will have to be seriously considered before English Agriculture can again hold the position it once held. I shall not be able to deal with it fully or philosophically; but I desire to draw public attention to this aspect of the question. We hear the cry everywhere that farmers are failing and that landlords cannot let their farms. It is generally admitted that British Agriculture is passing through a great crisis. The future of our Nation depends mainly upon the productions of the soil, it is therefore the simple duty of every Englishman to assist in solving the problem of the present distress. The subject is occupying, and will no doubt occupy many thoughtful minds. It is true that the seasons for several years past have been unfavourable, we have also had to contend with a keen foreign competition, that will make itself felt more and more for years to come. The weather we cannot control, although farmers may do much to modify the effects of unfavourable seasons by being well forward with their work, so as to be in a position to take advantage of favourable opportunities as they pass. But the foreign competition must be met, and grappled with, and that successfully, the question is, how and by what means it can be done, although free trade and protection will no doubt be seriously discussed in the immediate future, Agriculture cannot expect to find relief in this direction : something might be done by improved land laws, and, in some cases, a reduction in rents, although in many instances land brings in but a very poor return, and therefore very much relief cannot be expected even in this direction. My firm conviction is, the relief so much needed must be sought in internal rather than external remedies, we must apply ourselves, and that at once, to those remedies that are within our own reach, viz. : to economy, industry, skill, and sobriety. I am unwilling to believe that England is to be beaten by her foreign neighbours, although the contest is not equal (all round) in the matter of protection. I am persuaded, after long experience and much anxious thought on this subject, that by far the greatest hindrance to our Agriculture is England's common enemy or I might say curse, Intemperance, which has been literally forced upon the nation by the Government from age to age. I feel strongly, (speaking from a Christian's stand point) that had the Rulers intended to have made life miserable to millions of persons, to have produced crime, immorality, poverty, and eternal ruin to countless numbers of souls as well as the nation's ruin, no better method could have been adopted, than that of forcing upon the community licensed houses in such numbers and of such character as to have made it practically impossible to resist the temptation, and to avoid its really awful consequences, as may be seen in almost every parish throughout the land to-day. Whilst all interests in the country have suffered from Intemperance, perhaps none have been injured so much as agriculture. I will therefore refer to it in regard to its influence upon the various classes engaged therein, viz. :

1. The Landlords ;   2. The Tenants ;   3. The Labourers ;
and 4. Upon Agriculture itself.

I.   ITS BEARING UPON THE LANDLORDS.

Every landlord knows which has given him the most trouble, the intelligent, industrious, sober man, or the man that is neither the one nor the other. Who are the men that have to be excused at the rent audit ? Why landlords know, those as a rule who drink freely, men who have got into the habit of attending several markets a week, when perhaps one in a fortnight would be sufficient for business purposes, and thus a great deal of valuable time and money year after year is wasted.

Who is it that allows the farm buildings to become dilapidated and does not half cultivate the land, and gives up the property considerably deteriorated, which the landlord has to make good for the next tenant, either by expending a sum of money, or letting at a reduced rent for several years ?

The same is true as regards cottage property ; at least it was so upon an estate that I myself was connected with for six years, in Buckinghamshire. A cottage may be done up at the cost of say half a year's rent or more. It is let to an intemperate man and his family, in a very short time the place is all to pieces again. This is one reason why cottage property does not pay. I venture to think that the experience of most landlords who enter into the details of the management of their property will confirm my statement. Another serious burden that falls heavily upon the landlord is the increased rates and taxes through intemperance. If these matters have not been carefully noted and looked into, they may appear of but little consequence at any given time, yet if the records in any estate office for the last thirty years be carefully examined, they will be found to amount to a considerable sum.

The shocking condition of Ireland ought to be enough to make every English Landlord give this matter his most serious attention, for he knows not how soon a similar state of things may be at his own door. There can be no doubt but that the condition of that unhappy country is mainly due to the drinking habits of the people. According to the Government returns, the average rent is about 15s. per acre on 15,357,356 acres, which would make the rent roll of Ireland £11,518,392. Yet the average expenditure in drink by these starving people during the last ten years has been £13,823,102, a sum not only equal to the total amount paid for the whole of the land in the country, but exceeds it by £2,304,710. The people of Ireland have spent £138,231,020 during the last ten years upon this element, and this is only the direct cost, if we add the indirect loss through pauperism, crime, lunacy, loss of time, also loss of physical strength and mental power, we should have to add a great many more millions. Thus while the Duchess of Marlborough and others were collecting money in England to provide the people with food, &c., they were spending their money in this extravagant way—in drink—demoralizing themselves, and nerving their hands for dark deeds of cruelty and murder. Who can be surprised that people in such a condition should be a ready prey to agitation. I agree with the remarks of the Bishop of Cork, as reported in the *Times* a few days ago. "If the Irish would cultivate habits of Temperance, Industry, Thrift, and Self-Reliance, the evils of Ireland would disappear like the dew before the sun." Had the grand work of Father Mathew been sustained, Ireland would have been in a very different condition to-day. It is stated that through his efforts hundreds of thousands joined the ranks of the Temperance Societies, which had an immense influence for good in the country. The consumption of intoxicating liquors fell from 11,500,000 gallons per annum to 6,500,000 gallons, whilst aggravated crimes also fell considerably. If the Government and Landlords wish to find a remedy for Ireland's troubles, let them turn their attention to this question, by which they will do more for that unhappy country than by any other means.

If we take our own country, we find that more than twice the amount of the whole rent roll of the country for land, was spent in drink during the past year.* Can we be surprised that trade languishes, that agriculture is depressed, that discontent, distrust and poverty abound in the land. It could not be otherwise ; and who is to blame for all this ? not the poor unhappy victims, but the law. The Magistrates have given licenses without measure to the Brewers and others, and thus forced temptations upon the people, which they could not resist. The present state of things was clearly foreseen by Lord Chesterfield, who, speaking in the House of Lords 130 years ago, said—

"Drunkenness, my Lords, is universally, and in all circumstances, an *evil*, and therefore ought not to be taxed, but punished. The noble lord has been pleased kindly to inform us that the trade of distilling is very extensive, that it employs great numbers, and that they have arrived at exquisite skill : and therefore the trade of distilling is not to be discouraged ! Once more, my

---

* See the *Times* March 29, Leading Article—and Mr. W. Hoyle's Letter same date.

lords, allow me to wonder at the different conceptions of different understandings. It appears to me that, since the spirit which the distillers produce is allowed to enfeeble the limbs, vitiate the blood, pervert the heart, and obscure the intellect, the number of distillers should be no argument in their favour; for I never heard that a law against theft was repealed or delayed because thieves were numerous. It appears to me, my lords, that really if so formidable a body are confederate against the virtue or the lives of their fellow-citizens, it is time to put an end to the havoc, and to interpose, whilst it is yet in our power, to stop the destruction. So little, my lords, am I affected with the merit of that wonderful skill which distillers are said to have attained, that it is, in my opinion, no faculty of great use to mankind to prepare PALATABLE POISON ; nor shall I ever contribute my interest for the reprieve of a murderer because he has, by long practice, obtained great dexterity in his trade. If their liquors are so delicious that the people are tempted to their own destruction, let us at least, my lords, *secure* them from their fatal draught, by bursting the vials that contain them. *Let us crush at once these artists in human slaughter, who have reconciled their countrymen to sickness and ruin, and spread over the pitfalls of debauchery such a bait as cannot be resisted."*

This statement is confirmed by Charles Buxton, Esq., an English Brewer, and a Member of the House of Commons, who 100 years after in his essay, " How to stop Drunkenness," said—

"Not only does this vice produce all kinds of positive mischief, but it also has a negative effect of great importance. It is the mightiest of all the forces that clog the progress of good. It is in vain that every engine is set to work that philanthropy can devise, when those whom we seek to benefit are habitually tampering with their faculties of reason and will—soaking their brains with beer, or inflaming them with ardent spirits. The struggle of the school, and the library, and the church, all united against the beerhouse and the gin palace, is but one development of the war between heaven and hell. It is, in short, intoxication that fills our jails ; it is intoxication that fills our lunatic asylums ; and it is intoxication that fills our workhouses with poor. Were it not for this one cause, pauperism would be nearly extinguished in England.

Looking, then, at the manifold and frightful evils that spring from drunkenness, we think we were justified in saying that it is the most dreadful of all the ills that afflict the British Isles. We are convinced that, if a statesman who heartily wished to do the utmost possible good to his country were thoughtfully to enquire which of the topics of the day deserved the most intense force of his attention, the true reply, the reply which would be exacted by full deliberation, would be that he should study the means by which this worst of plagues can be stayed. The intellectual, the moral, and the religious welfare of our people, their material comforts, their domestic happiness, are all involved. The question is, whether millions of our countrymen should be helped to become happier and wiser—whether pauperism, lunacy, disease and crime shall be diminished—whether multitudes of men, women and children shall be aided to escape from utter ruin of body and soul ? Surely such a question as this, enclosing within its limits consequences so momentous, ought to be weighed with earnest thought by all our patriots."

These men cannot be considered biased or fanatics. No, their statements are true to the letter, as experience has shown that the liquor traffic is the enemy of God, of man, and of all social and religious progress. It has always been a mystery to me why the Bishops and Clergy, the Landlords and the upper classes generally should have been so opposed to the great temperance cause. Whether they know it or not, they owe much to it. One of themselves has said "The more I examine and travel over the surface of England, the more I see the absolute and indispensable necessity of Temperance Associations. I am satisfied that unless they existed we should be immersed in such an ocean of immorality, violence and sin as would make this country uninhabitable."—*Lord Shaftesbury's Speech at Norwich.*

II. ITS BEARING UPON THE TENANTS.

I fear it may give offence to some of my readers when I say that farmers have been great sufferers from intemperance. I don't mean from the vulgar drunkenness of the sot, but from free and heavy drinking, which has led to a serious waste of valuable time and money. I have travelled in many parts of the country during the last 25 years, and paid special attention to this question, and I am obliged to arrive at the above conclusion, and from the nature of things I do not see how it could be otherwise, for till recent years we were all taught to drink, it was the one thing that special attention was given to, we were told that drink, above every other thing, was what we needed to make us grow, to make us strong, to keep us in health, to enable us to enjoy life, nothing could be done without it, no bargain could be made without drink, when and wherever friends met, they must drink if it was within their reach ; when the farmer went to see his landlord, drink was offered him, at the rent audit it occupied the chief place at the feast, when he went to market he was surrounded by inducements to drink. There can be no doubt but that thousands of men can trace their ruin to the temptations of the market. I have seen families brought to ruin and have watched with sorrow the process. In some cases it has gone on for ten, fifteen, twenty, or more years. I feel sure that the reader, if he reflects for a moment will be able to call to mind many cases within his own experience. The process is as a rule gradual, but sure ; and many know it, but few have the the moral courage to cut themselves off from it, which is their only safety.

It may first begin with the social glass at home, or the market glass with friend after friend, and by degrees the drink craving is created : it calls for a glass in the morning, it reminds its victim when he is passing the house where it can be had, the process goes on till the craving becomes a vice, and the vice a disease; and now he is fairly on the way to ruin, time and money are increasingly wasted, self respect is gradually but surely lost, the intellect becomes clouded and weakened, the moral character of the man is destroyed and its effects is visible in the family, and from this point the work of ruin is soon completed either by distress or death. In almost every village in the country, in this way drink is doing its deadly work. It is not always the head of the family; wives, sons, and even daughters fall victims to this vice. It may be thought that I am over drawing the picture, which I would not do on any account if I knew it. The evil has existed long, and is deeply rooted. I would advise any of my readers to put this statement to the test in their own parishes. Let them take say the last 30 years, and take each house, and see what has happened, to each family, from the effects of drink, and I venture to believe, that their eyes will be opened then, if they have never been before, to the bearing of Intemperance upon Agriculture. I have recently done this in connection with other persons in several villages in different parts of the country with sad results.

A farmer lately failed in one of the Midland Counties, whose liabilities for drink and tobacco amounted to £1000. Another case in the same neighbourhood : a farmer stated to a friend of mine that he had paid over £300 in one year for drink ; these I grant are extreme cases, yet it must be admitted by all acquainted with the habits and customs of farmers, that a vast amount of money is too often spent upon drink and its associations.

There can be no doubt that the worst enemy all down the ages to British agriculture, has been, and is, the drink. It not only empties the pocket, but it seriously weakens the brain-power, and prevents the cultivator from grasping, and grappling effectually with the difficulties that beset his path. Lord Salisbury is reported to have said in a speech a short time ago, " That what was most needed in English Agriculture, was brain-power." This all must admit to be true, and what weakens the brain and deadens the intellect so much as intoxicating, or, more correctly—poisoning drinks. As shown by the following statement of Sir William Gull, before the select committee of the House of Lords, July, 1877—" I should say from my experience, that it (alcohol) is the most destructive agent that we are aware of in this country. I think that, taking it as a whole, there is a great deal of injury done to health by the habitual use of wines in their various kinds, and alcohol in its various shapes, even in so-called moderate quantities. And to people who are not in the least intemperate, I think drinking leads to the degeneration of tissues; it spoils the health and it spoils the *intellect.* I know it (alcohol) is a most deleterious poison. I would like to say that a very large number of people in society are dying day by day, poisoned by alcohol, but not supposed to be poisoned by it."

I have never known a man fail that farmed on Temperance principles, but I have known several labouring men, by those principles, through God's blessing, raise themselves, and become successful farmers.

Farmers, like other employers, lose through the intemperate habits of their men, by loss of time, accidents, inferior work, and in many other ways, such as poor rates, county and school board rates, in having to mantain the large prisons, lunatic asylums, workhouses, industrial schools, &c. This is also true to a large extent, as regards hospitals and a large number of other institutions, whose existence is mainly due to intemperance, and have to be maintained at a great cost by public subscriptions. Taking all these together, they constitute a serious burden upon the land, and if we can banish intemperance from our midst, four-fifths of this burden would disappear. The Rev. W. L. Blackley said at a conference in the Sheldonian Theatre, Oxford, November 6th, 1880 :—" My parish has a population of about 400 souls, an annual poor rate for the support of the poor only, of £150, and it contains three public houses, and one beer shop, supported, as it happens, exclusively by the agricultural class. I assume that to keep a landlord and his family, and to pay rent and license, we must allow an average gain of at least thirty shillings a week, and, supposing that for the drink which gives him this profit only to cost even thirty shillings, it becomes clear that to keep up the drink trade in the four shops in my parish, costs £12 a week at the very least, or an aggregate of £624 a year, more than four times the poor rate required to support our pauperism. Now, who are they who become chargeable to poor rate ? The victims of the drinking customs and drinking habits to-day, and the drinkers themselves by and by. If the labouring class (and there are no other public-house customers) in my parish, can afford to spend a pound in drink for every five shillings the rate-payers are compelled to contribute for the support of those whom drink ruins, it is clearly plain that, were the drink traffic in general reduced by even 25 per cent., there would be no need of poor rates at all, and, were that traffic abolishsd altogether, each father of a family in my parish would have enough money in ten years time, to buy himself, if he thought it necessary to his sense of independence, the fee simple of two acres of the best land in the place, and thus to solve the question once for all, of peasant proprietorship of land in the only way possible, except by the exercise of spoliation at the risk of revolution."

Then again the farmers suffer in consequence of not being able to find a ready sale for their produce. I state this on the authority of Col. Lloyd Lindsay, M.P. for Berkshire, who is reported to have said in a public speech—"That the British farmer wanted a quicker sale for his produce, or in other words more purchasing power was wanted, but the people did not possess that power." I maintain that the Hon. Gentleman was mistaken, the people do possess that power if it was spent in needful food and useful articles of clothing, furniture, &c., as is abundantly shown in Mr. Blackley's statement above. Can we be surprised that not only agriculture, but that trade and commerce generally should be depressed, when we think of the wicked waste of wealth spent in purchasing intoxicating drink, a crime, poverty, and disease-producing element? Take for example the year 1878. Over £142,000,000 was spent in that one article, an amount equal to 8s. per week for each family, or £4 4s. per annum for each man, woman and child in the country. These figures are taken from the government returns, but they by no means represent the total expenditure, for I need not attempt to prove that spirits and other drinks are considerably added to after they leave the manufacturers. Then if we add to this the indirect loss, which is considered by the best authorities to be almost as much as the amount spent, making the total £284,000,000, considerably more than all the foreign trade of the country. Mr. Caird in his work entitled "The Landed Interest" (see page 14) shows that the entire value of all the produce of all the land in the country is only about £335,000,000, or only about £50,000,000, more than the loss caused to the country by its drinking. If we take the seven years, from 1871 to 1877, we get the frightful amount of £964,037,836 spent in drink, £210,000,000 more than the whole of the national debt. During that same period 3,334,110 persons were convicted of crime, and £101,144,718 was paid in poor and police rates, and in that some period 1,271,838 drunkards were apprehended. During 1879, 3,000,000 persons were upon the books of the Poor Law Unions, and it has been stated before the Social Science Congress and the British Medical Association, by Dr. Norman Kerr, that about 120,000 persons lose their lives in this country annually through drink, either directly or indirectly. Was there ever such a state of things as this in any country in the world before? May we not adopt the language of the prophet and say "The land is full of bloody crime and the city is full of violence." With these facts before us we cannot be surprised that agriculture is depressed, that purchasing power is weak, trade generally bad, and that we are not able to compete with our foreign neighbours. No doubt much harm has been done to trade by strikes and contentions between capital and labour of late years, but I repeat that the drink system is mainly, if not wholly, responsible for all this. Men will strike for two or three shillings more money per week, at the risk of ruining a firm and driving trade from the country, while they think nothing of spending two, three, or perhaps four or five times that amount in drink. Men in this condition are easily led by any agitation that may arise, with a plausible idea of doing something to better their position, and we cannot wonder at this when we think of the miserable poverty-stricken homes they live in. But on the other hand had they practised habits of temperance during the last 25 years, they might have had millions of money at their command, and the strikes and their consequences would never have taken place : good feelings would have existed between employer and employed. In proof of this I quote the following valuable testimony which I take from the " Scientific American," " Hon. W. E. Dodge stated that his firm employed 2,000 hands, and made it a rule that their people should not use intoxicating drinks- They had no complaints of hard times among them, they had been able to stand against the depression, the men had accepted such wages as they were able to pay, and many of them owned their own houses—and that crime was practically unknown among them."

I earnestly commend this example and its results to English employers and men. The employers ought to lead the way, as a rule they have taken but little interest in this movement. The farmers generally speaking have treated all temperance efforts with contempt. Such conduct is unworthy of so important a class of the community. I know that some employers say that they get more work done by giving the men beer in their work, others say that they cannot get men to work without it, but these are mistakes, as all experience shows. Firmness and determination and giving an equivalent is all that is needed.

The practice of giving beer in the hay and harvest field is a remnant of the old truck system which the law has now abolished. There can be no doubt that this practice has had a very bad effect upon the men and their families, and it ought not to be continued. I will not dwell at length upon this aspect of the question. I have collected a considerable amount of evidence during the last few years from employers of high position in various parts of the country, all showing the undeniable advantages of not giving beer in farm work. Some of these valuable testimonies will be found at the end of this pamphlet, which I would earnestly commend to the serious attention of all classes. There is everything to encourage both masters and men to make the experiment. Prejudice, I fear, has kept many from adopting this needful reform, but I feel sure that it must soon give way in this case as it has done in the past. The prejudice of many employers and men was very strong, when some advanced minds began to plough with a pair of horses abreast some 60 or 70 years ago. The same prejudice existed against winnowing machines, haymaking machines, and in fact machinery generally, but time has shown that it was mistaken prejudice, and I have no doubt but that a little time and patience will show that the opinion given by Mr. Clare Sewell Read, ex-M.P. for South Norfolk, "that a wet groat is better than a dry shilling in hay or harvest work" was also the result of mistaken prejudice or want of knowledge or experience on the subject. The practice ought to be discontinued;

it is alike injurious to employers and men, it is morally
financially, and physically wrong. After reading the valuable testimony in the appendix, no honest minded man can
fail to see that beer does not help the labourer to do his work
but is a positive hindrance, what we want is to get rid of the
idea of either beer or beer money, as regards payment for work,
both in the field and in the house.

III. INTEMPERANCE, ITS BEARING UPON THE LABOURERS.

Intemperance has borne heavily upon the working
classes of England, and the farm labourer has had his share
of the burden. From his infant days he too is taught to believe
that beer is the one thing needful for this life. He is told that
it will make him grow, give him strength to do his work, will
comfort him in his trouble, give him health in sickness; in fact,
that it will be his friend at all times, and under all circumstances;
when he leaves his home his education in drinking is continued
by his employers who give it to him, encourage him to drink it,
and in many cases, long before he is able to think for himself on
life and all its great realities, the drink craving has got hold of
him, the vice is acquired, and perhaps the organs of his body are
more or less diseased; he is an unconscious victim, and far on
the road to ruin, by the time he arrives at a mature age. In this
state he marries, with little or no means to make his home comfortable; in poverty he sets up his home, he continues to drink,
his family increases, his home is miserable, little furniture, little
food, little clothing; children badly brought up, who, like
their father, in too many cases, fall into the national vice, and the
evil is perpetuated from generation to generation with all its
horrors. What a happy contrast the homes of the English working classes would have presented, had they only been able to have
resisted this vice, and have practised habits of temperance and thrift.
What an unspeakable blessing to themselves and the nation.
There would have been no fierce conflict between capital and
labour. Unions would not have been heard of; peace, plenty,
contentment, morality and religion, by the blessing of God, would
have abounded throughout our land; but, alas, the very reverse
is true? It need not be so; it is not so much for want of means,
because the income of the agricultural labourer who earns the lowest
wages of any class in the country, has increased nearly 50 per cent.
during the last half century. I will mention a fact that came
under my own notice upon the estate that I was connected with
in Buckinghamshire. The home farm was under 500 acres, the
earnings per week for 15 men in 1874 was as follows: one, the
hay binder, who worked early and late, 24/-, four over 19/6,
three over 18/6, two over 17/6, three over 16/6, two over 15/3;
and in addition to this, some of their wives and families earned a
considerable sum, and yet through the money wasted in drink
some of these people were living in a most miserable condition.

£60 was paid for the beer given to the men upon that farm the same year, which I need not say,—notwithstanding the greatest possible care,—had a very demoralizing effect upon them, some of them leaving the farm night by night the worse for drink. If we add to the £60 worth of beer given by the employer, the moderate sum of 1/6 per week spent by themselves in drink and tobacco, it would amount to about £8 for each of the 15 families; and if laid by year by year at 5 per cent. compound interest, it would in ten years amount to over £100 for each family. I need not say more to prove that intemperance has a very terrible bearing upon the physical, social, moral, and religious welfare of the English agricultural labourer and his family. I trust the time is at hand when all employers will feel it to be their duty to pay in cash instead of giving beer, especially to the boys, and at the same time do all they can to encourage habits of Temperance and Thrift.

IV. INTEMPERANCE, ITS BEARING UPON AGRICULTURE.

The reader will have seen by this time that intemperance has a very important bearing upon Agriculture. He will see, (1), how it affects the Landlord in relation to his property, in the increase of his rates and taxes, and in a variety of other ways. (2), how it affects the Tenant, in the loss of time, capital, and mental power, in the increased rates and taxes, trouble, accidents, neglect, inferior labour, and other losses through intemperance. How it affects the markets and prevents his meeting with a ready sale for his produce. How this would be altered, if say £50,000,000 of the money now spent in drink could be spent in the productions of the farm. (3), how seriously it affects the poor labouring man and his family, how the interests of landlord, tenant and labourer are bound up together, and what affects one affects all, more or less. Some farmers may say, yes, it is all very well from your point of view, but if we become a temperate nation, what shall we do with our barley? This is a matter of some importance, but it is a narrow and somewhat selfish view to take, it is probable that less and less malt will be used year by year, since the tax has been taken off malt, and put upon the beer. Sugar, maize, rice, and other things are already being used extensively. Between 1878 and 1879 the consumption of malt decreased 7,341,812 bushels; between the same dates the quantity of sugar used in brewing increased 538,461 cwt., and moreover I am persuaded that the practical farmer will be able to dispose of all the barley it is absolutely needful for him to grow, for feeding and other purposes, and the small loss he might sustain by not being able to sell his barley for malting, would be more than balanced by the increased demand for his other productions.

"If the people of this country are poor, the only customers of the farmer are poor, and a poor man cannot be much of a customer. Enlightened self-interest, therefore, ought to make the farmer, and above all, the landlord, a persistent enemy of intemperance. Some people still fancy that the drink trade helps the 'landed interest,' because it consumes a large quantity of barley. The drink trade is a big customer : yes, but the nation is a much bigger one ; and considering the damage which the drink trade does to the nation, and considering the extent to which it impairs the purchasing power of great masses of the people, the few millions obtained for barley are but a trifle compared with the loss caused by intemperance."—*Alliance News.*

When railways were introduced, the question was asked, what shall we do with our horses? That problem is solved without any injury to the farmer. With good management, high cultivation and favourable seasons, much may be done to contend successfully with foreign competition, not only in the production of beef, mutton, pork, poultry, butter, cheese, eggs, &c., but in grain also, and depend upon it, the secret of success in Agriculture, as of other branches of trade and commerce, is to be found to a considerable extent within the four corners of this question.

AN APPEAL TO THE CLERGY, LANDLORDS, EMPLOYERS AND LABOURERS.

I would again earnestly appeal to all landlords, employers and labourers, to give this great and most important subject their serious consideration. Landlords can help by doing away with the sale of intoxicating liquors in their villages, and in many other ways. When lately passing through the village of Everingham in the east riding of Yorkshire, I was delighted to find that Lord Herries had set such a good example in this respect. His Lordship, on seeing that harm was being done to the people in the village through the drinking going on at the public house, did away with the license, and made the house into a Temperance Inn, and, in order that the publican might be able to get a comfortable living, his Lordship attached a little more land to the house. Farmers can help by at once adopting the plan of paying in cash instead of giving the men beer, and encouraging temperance by their own examples and efforts. Men can help by thankfully receiving the extra money, making good use of it, and practising habits of temperance themselves, and encouraging the same in their own families. But above all I would appeal to the Bishops and Clergy of our church, for it is only when the church wakes up to a sense of her duty in this matter, that we may hope to meet with success. I do feel most strongly that till within the last few years, the church has lamentably failed, in the discharge of her duty to the people in this respect, not only have the Clergy as a body, (with but few exceptions,) stood by, looking on with cold indifference, not only content with making no special effort in this direction, to save the masses perishing around them, but in many instances openly opposing the efforts of others. Some have even refused the use of the parish schoolroom, to bodies of working men, who were anxious to associate together for mutual protection of themselves and families against the common enemy that threatened their destruction. Although a loyal and whole hearted member of the church, I do feel that the past conduct of a large majority of the Bishops and Clergy in reference to this question of intemperance is indeed much to be regretted. After taking a lively interest in the common welfare of my fellow working men, for upwards of a quarter of a century, and observing the effects of the drink curse which has been forced upon them, producing, as it has, the truly awful amount of poverty, misery, crime, cruelties, murders, disease and sending 60,000 or 70,000 human beings to a drunkard's grave year by year, and to think of the bearing of all this upon their eternal condition, viewing it in the light of the New Testament, and the teaching of the book of Common Prayer, it overwhelms my soul. This brutalizing system on the one hand, the neglect of the church on the other, are the main causes of the masses being forced into heathenism or even something worse, for not only immorality, but infidelity, is rife throughout the country, and how could we expect any other result? When preaching recently at St. Andrew's Church, Ancoats, the Bishop of Manchester stated "that in a Manchester parish containing 1,233 houses, the clergyman found as the result of personal inquires, that the heads of 907 families openly professed that neither they nor their households attended any place of worship. Ninety-three families called themselves Church of England people, 94 families called themselves Roman Catholics, and the rest were made up of different denominations, the Wesleyans being strongest with 54 families. The fact that 906 families out of 1,233 never attended public worship, was, the Bishop remarked, a scandal and a peril to society."

A justice of the peace for the county of Berks stated at a public meeting at Wallingford, "that he had been a great employer of labour abroad, and had travelled a great deal, and was sorry to have to come to the conclusion that the English working classes were the most degraded people he had ever met with." We are worse than the heathen nations. Sir Rutherford Alcock recently said " He could conscientiously aver that he saw more degradation, violence, misery and brutality in a single day in the streets of London than of twenty years of life in China." A Clergyman, who had spent many years in heathen lands, told me he found on his return that he had left civilization behind and come home to barbarism. What a disgrace to our nation, with an established Church professing to provide the needful means of grace, with a school in every parish, and during the last quarter-of-a-century—with the exception of the last few years—our country has passed through such a period of prosperity as few countries in the world ever did, and yet we are in this lamentable condition. In fact it has counteracted to a great extent the effects of the gospel of Christ and proved to the world that Christianity in England, in the 19th century, has failed to grapple effectually with immorality and sin. Within the last few days at a village I have known for many years, I asked one of the oldest and most respectable residents, who has known the place well for fifty or sixty years, to tell me who, to his knowledge, had suffered through intemperance during the last 30 years. We sat down together and carefully went throughout the whole parish, house by house, and out of 86 families we found that 97 persons had visibly suffered through intemperance; some ruined, others brought to premature graves, one killed while drunk. One poor woman, the wife of a farmer said in her dying hour, "my life has been a life of sorrow and trouble." Yes, God only knows the agony of that woman's soul, and the floods of bitter tears she shed during the long number of years she was tied to that husband with her six or seven children. During the 30 years above mentioned there had been 4 incumbents in the parish, but not one of them had taken any steps to inform the people of the nature, effects and character of alcoholic drinks, or used any special means to stay its ravages. And is not this parish, both as regards the evil effects of drink and the neglect of the Church, but a sample of thousands of others throughout the country? I would earnestly ask those clergy—who have hitherto stood aloof from or in the way of temperance work—in the name of the suffering, perishing masses, to stand in the way no longer, but at least to give the people the opportunity of understanding this question—especially in villages —by sermons, lectures, literature, &c. If they will not do this, surely the blood of the slain will one day be required at their hands. This is not only a moral but a physical question. Intemperance puts a physical impediment in the way of a spiritual impression. How unutterably sad it is to see so many of the Ministers of Christ's Gospel to a fallen world, stand by in the present distress as if they had no heart to feel for or hand to help a fallen brother; and I believe the chief causes are prejudice and want of knowlege of the subject—the first ought never to have existed, the second could have been removed by a few hours of careful study. Who can tell the fearful amount of human suffering that would have been prevented, and the myriads of precious souls for whom Christ died have been saved if the great body of the Clergy had but done their duty in this respect. I again appeal to them, in the name of all that is good, to come out like men and take their places right in the front ranks of the Temperance army as Captains of tens, of hundreds, or of thousands, in their various parishes, and the day has now come when they must do it, the state of the country imperatively demands it of them, for the shocking state of things mentioned by the Bishop of Manchester is but too true a picture of the county generally as shown by the recent religious census. What a subject to reflect upon for the Bishops and 20,000 clergy. Would it not have been better for millions of English men and women to have been born in the darkest corner of heathen Africa, where neither Christianity nor civilization had ever been heard of, and why? Not because Christianity and civilization are not the greatest of blessings, but because the conditions under which they live are such as to make it practically impossible for them to benefit by the advantages that would otherwise be of the greatest possible blessing to them, and at the same time they are responsible for these advantages. Some will no doubt say that the drink is God's good creature, and on that ground justify the existing temptations; but suppose a shepherd, entrusted with his master's flock, was to allow a number of wolves to establish themselves among the sheep, and when the master, who, seeing that his flock was being destroyed, called him to account for his conduct, the man was to say, "O, Sir, don't speak in that way of the wolves, I felt I ought to allow them to remain in the fold, because yon know, Sir, they are God's good creatures." I think the injured master would say, "Yes they may be God's good creatures, but they are out of their place in my sheep fold." What he would do with the shepherd I will leave the reader to judge; but one thing I am certain of, that he would say it was hopeless to attempt to rear a flock under such conditions. Others will say, "Preach the Gospel." Yes, very good, but has not that been done, and what are the results as far as the masses are concerned, and why is it so? Is it that the Gospel has lost its power. Certainly not. The real reason is that there is a mighty man-placed hindrance in the way. The masses are buried beneath the demoralizing influence of the liquor traffic, fast bound by the chains of their sins. There is the stone to be taken away, and the grave clothes to be loosed, and for ages the Lord God Almighty has been calling upon His church to "take away the stone," that He may give newness of life to the dead in sin; but the church has turned a deaf ear to the heavenly voice, and has not until lately rebuked the national sin, but has been, to some extent, an apology for its existence, and allowed it to find a refuge under her shadow.

I believe that myriads of the working classes would be thankful to be delivered from the heavy burdens caused by the drink, if they had the opportunity and needful assistance given them. I am justified in saying this, by the tens of thousands that have lately joined the Salvation Army in various parts of the country—which is the stones crying out because the Church has held her peace, and passed by on the other side.—It is reported that in Leeds alone 25,000 joined; and none are accepted unless they are prepared to give up the use of strong drink and tobacco. The same desire was manifested on a large scale in Ireland during the days of Father Mathew. Numbers of earnest working men and others have been struggling for the last 50 years to deliver themselves and their fellows from this curse, worse than Egyptian bondage or American slavery, but they have had to contend against long odds. The upper classes and employers generally gave the movement the cold shoulder. The press, to its shame, till lately opposed their efforts. The Doctors, as a rule, were opposed, and did much harm by ordering their patients drink right and left; and, above all, the Church of God gave a deaf ear to the cry for help;—but, thank God, a brighter day appears to be breaking upon us at last. Some of the Bishops are waking up to a sense of their duty in this matter, and an increasing number of the Clergy are manfully grappling with the foe. The Doctors are yielding to the force of truth, and the educated classes are giving serious attention to this movement. The press is also realizing its importance—the old proverb is again being verified "That truth is great and will prevail," but the battle will be fierce and difficult. The enemy will die hard ; we shall need to work with all our might and pray with all our heart that God will be pleased to give the nation a speedy deliverance. I would therefore invite all who are so disposed to join in a common petition to the throne of God day by day, and for that purpose append the following form of words which are used by the Members of the Church of England Temperance Society's Prayer Union :—

"O God, whose blessed Son was manifested that He might destroy the works of the devil, and make us the children of God and heirs of everlasting life, look down in mercy upon all those who are engaged in Thy name in combating the deadly sin of intemperance. We confess, O Lord, that in the time past of our lives we have sinned against Heaven and before Thee, in the indifference we have shown to the progress of this terrible evil among us. We own, with deep shame and humiliation, that by reason of the abounding drunkenness in our land, cruelties have been perpetrated in countless numbers, and of the deepest dye ; the sacred love of family has been torn and riven asunder ; the sorrowful sighing of women and children, by thousands and tens of thousands, has gone up and entered into Thine ear ; even now the land is defiled with the blood which the drink is continually shedding ; and while in our own lands souls are kept in bondage and are perishing with none to help, through our fault the name of Christ has come to be largely blasphemed among the Heathen. It is because Thy mercies fail not, but are new every morning, that we have not long since been consumed. But, O Lord, we repent of these our misdoings ; we pray Thee to give true repentance to ourselves, and all the people whom Thou hast called by Thy name. And now that we are associating, in Thy name, to meet the evil with special remedies, grant that Thy Holy Spirit may be with us to direct and prosper us. Be to us, O Lord, a Spirit of wisdom in all our councils, a Spirit of unity and brotherly love in our several associations, a Spirit of ghostly strength in the enterprises to which we may address ourselves ; so that casting away all self-confidence, and depending upon Thy most mighty protection, we may triumph over all the opposing work of the enemy. Give us the grace of patience under trial, of perseverance under every seeming want of success, of meekness under reproach, of endurance under temptations, whether of the flesh or spirit, and above all, of singleness of aim and motive, that in all we do we may seek Thy glory and the good of our fellow-men. Shed abroad in our hearts the love for our perishing brethren, that we may be willing for their sakes to deny ourselves in lawful indulgences, if by so doing we may save some of them. And because the hearts of men are in Thy control, to turn them as it seemeth best unto Thy godly wisdom, take away from among us, O Lord, all blindness and prejudice, and whatever else may hinder the advance of the cause we have at heart, so that the number of faithful hands and earnest hearts being continually increased, we may be at length permitted to attain to the end at which we aim, a Temperance Reformation in our beloved land ; grant this O Heavenly Father, for Jesus Christ's sake, our only Lord and Saviour. Amen."

APPENDIX.

* TEA v. BEER IN THE HARVEST FIELD.

A well-attended Conference of the Members of the Newbury Chamber of Agriculture was held on July 25, 1878. in the Town Hall, Newbury. to consider " How far it was advisable and practicable to introduce non-intoxicating drinks into the Harvest Field."

The chair was occupied by the President of the Chamber (Mr. E. J Deverell).

T. BLAND GARLAND, Esq., J.P., Burghfield, Reading, said he had recently, at the request of the Oxford Diocesan Secretary of the Church of England Temperance Society, written out his experience on the subject now before the Chamber, which was as follows :—

"Previous to 1871, in common with all other employers in this neighbourhood, I had been in the habit of giving my labourers beer during the hay and corn harvest ; after much reflection I came to the conclusion that the practice was quite unjustifiable, that it was worse than the very worst form of the " truck system," as it did not even give the labourer the option of selecting the articles which he received in part payment of wages, but obliged him to receive, and practically to consume, an expensive and unnecessary article, in such quantities as to be prejudicial to his health and the well-being of his family ; in other words, that the practice amounted to expending about a fourth of his wages, without his consent, in the purchase of beer, and obliging him to drink it. Employers sometimes say, " We do not give the beer to our people in part payment of wages, but as a gift in addition to wages ; and we do not oblige them to drink it,—we only offer it ; they may take it or leave it." These are mere idle evasions. If an employer pays his labourer 2s. 6d. per day in money, and expends 1s. more in purchasing beer for him, clearly he estimates the value of his labour at 3s. 6d. per day, and pays that sum for it. Again, if he employs a number of people in thirst-creating labour, such as that of the harvest field, and supplies them exclusively with a drink which they have been educated to prefer, he practically obliges them to drink it. No better system could be devised for insuring the drunkenness and poverty of our agricultural labourers. As mere children, they are tempted to drink excessive quantities of beer, because it appears to cost them nothing, and they learn to consider the quantity consumed as a test of manliness.

It is no exaggeration to say that in those districts where beer is supplied by employers in the harvest field, its cost, and the further amount expended in its purchase by the labourers in consequence of the taste for it thus acquired, would suffice amply, with careful investment, to provide for the comfortable maintenance in sickness and old age of the whole agricultural population.

I maintain most positively that nothing can be more unsuitable as a thirst-quenching beverage during hard work in hot weather than beer. Even at the best it is an unnecessary luxury, which our agricultural labourers cannot afford, and I submit that their employers do them a grievous harm by encouraging them to use it.

In 1871 I determined to supply no more beer to my labourers under any circumstances ; and I agreed with them, as an alternative, to pay the men

---

* Those marked thus are published as Leaflets by the Church of England Temperance Society.

18s. per week instead of 14s., and the women 9s. per week instead of 7s., during the hay and corn harvest.

After this was settled, reflecting that the people would probably find it impossible to supply themselves with a suitable substitute for the beer, and would in a measure be driven to the public-house, I determined to supply them with tea. I therefore purchased a common flat-bottomed 8½ gallon iron boiler, with a lid, long spout, and tap; this is taken in a cart to thě field, with a few bricks to form a temporary fireplace, a few sticks for the fire, some tea in 7-oz. packets, and sugar in 4-lb. packets. The first thing in the morning, a woman lights the fire, boils the water, the bailiff puts in the 7 ozs. of tea in a small bag, to boil for ten to fifteen minutes, then removes it and puts in 4 lbs. of sugar; if skim milk can be spared, two to four quarts are added, but this is not a necessity although desirable. All the labourers are then at liberty to take as much as they like at all times of the day, beginning at breakfast-time, and ending when they leave off work at night. If the field is large, they send large cans to the boiler for it ; so soon as the quantity in the boiler is reduced to two gallons, it is drawn off in a pail for consumption, whilst another boilerful is being prepared. The knowledge that they have at their disposal as much good tea as they choose to drink during every minute of the day, materially lessens their thirst.

This system has given the most perfect satisfaction during the last eight years : the people have done more work with comfort to themselves, have never been stupefied and quarrelsome, which they always were, more or less, in the beer days, and go home sober at night instead of to the public-house, as many formerly did.

The increase of wages to be paid must of course depend on the quantity of beer usually consumed ; this varies in different localities ; but I have found that a careful examination generally proves it to have been much more than the employers suppose it to be.

The cost of tea in my case is as follows :—

	s.	d
7 oz. of tea	1	0
4 lbs. of sugar	1	2
Skim milk, about	0	2
	2	4

or 8½ gallons of tea, at 3½d per gallon.

I had 28 men and women employed in hay-making this year, and the consumption was—

		gals.
Generally,	2 boilers full per day	17
Occasionally,	2½ ., .,	21¼
On one day,	3 ,, ..	25½

My calculation is, that they drink on the average two-thirds of a gallon each per day, at a cost of 2d.

Thus I pay them in lieu of beer 8d. per day in money and 2d. in tea, or 10d. in all.

But if the change involved a much larger expenditure than the cost of the beer, employers would be amply remunerated in the better and larger amount of work done, the better disposition of their labourers, the decrease of pauperism, and the general well-being of the people.

As a last word to employers. I would say, no half measures; let the additional wages be given to the full value of the beer ; let the tea be good, and made with care in the field, not sent out from the house, or there will not be enough ; be sure that it is always within the reach of every labourer, without stint. See to this yourself: trust it to no one ; beer has many friends. Be firm in carrying out the change. and it will be a source of great satisfaction to you and to your labourers, with very little trouble and at no extra expense."

After two hours' discussion of the subject, the following resolution was unanimously adopted :—

" That this Chamber considers it desirable to lessen the quantity of beer carried into the harvest field. by paying in money rather than in kind ; and that it is also well to endeavour to provide non-intoxicating drinks, to prevent the men spending their money at public-houses."

HARVEST WORK WITHOUT INTOXICANTS.

Leighton Park, Reading,
June 11th, 1880.

* Dear Sir,—Some months ago you asked me to give you an account as to how I was able to get on at harvest work without beer, and as the hay-making time has again come round. perhaps this will be the most appropriate season to furnish you with the information. although I must apologise for not having replied to your inquires before.

About two years ago, after reading a letter of Mr. Bland-Garland's, of Hillfields, in this county, on the subject of "Tea *versus* Beer," which appeared in the Church of England Temperance Chronicle, my mind was made up to discontinue the beer allowance, which up to that time I had made to my labouring people during hay and corn harvest; and I have no hesitation in saying that, since I gave up the practice. these operations have been carried on more satisfactorily in every way : certainly so to my bailiff, and I believe also to my labouring people themselves, who seem to prefer the money, and work better and more contentedly. I have since been more free from the many petty annoyances so generally caused by the state of independent swagger which an allowance of a pint of beer every two hours during the day is apt to promote, and for which those who give it have to thank themselves, It is done under the impression that the labouring people will have it because it is the custom, and that there is no use in trying to fight against the established usages of the country. But these are what the cause of temperance—when the customs are bad—has most resolutely to contend with ; and in many cases a moderately bold front will show how very mistaken the idea is that such difficulties cannot be overcome. The best proof I can give you of this, as regards the harvest beer question, is the answer in his own words, which one of my best labourers, a hale, strong man, gave to the question : How he liked the change from beer, to beer money ? He said, " I should be a fool not to like ; for I can do very well with two pints of beer a day, which leaves me five pence in money ; and I get tea into the bargain," and as far as I can judge many of my harvesters limit themselves to about this quantity. Those who cannot, or think they cannot, do without it, bring it with them, or procure it at meal times ; but they do not, as some fear they would, absent themselves continually from work for the purpose of getting beer.

Many farmers in this neighbourhood give money allowance of one shilling a day in lieu of beer ; and I was surprised to find, when I came to inquire, how many did so, but I prefer, and adopt the plan of giving 9d a day in money and as much tea as they can drink. My reason for doing so is that it is an educating process. Many labouring people say, and no doubt believe, that they cannot do harvest work without the large quantity of beer which is generally given them, but when the tea is put before them, they soon learn that beer is not the indispensable article they had considered it to be, and although it may still be indulged in as a luxury, it is found not to be a necessity.

The details of my plan are simple. I began by purchasing in Reading, what Mr. Shackleford, the tradesman from whom I bought it, described as a strong eleven-gallon boiler, with copper flue complete. at a cost of £3 15s ; some more rough and ready kind of boiler at less cost might, no doubt, be procured to do the work quite. or nearly as well as mine ; but, having made a vicarious purchase, I am proud of my teapot, and often on a hot day, find myself inclining towards it for a drink.

Our brew of eleven gallons is made at a cost of about 2s.6d., as follows : half a pound of tea at 2s. 6d., 1s. 3d. ; four pounds of sugar at 3½d., 1s. 2d. ; milk, about 1d. Total 2s. 6d. Milk is kept on the ground, and added when required ; about half of my people prefer it without. The number usually provided for is about twenty-two, and they consume about half-a-gallon a-piece during the day, but the last two seasons have been cool, and in a hot one it would probably be more. The first brew is made early in the morning, and is generally enough : but if more is wanted it is made, for it is essential to the success of the plan that tea should always be forthcoming when asked for. It is placed in the most central part of the ground, always ready to fill the tins or the bottles that are brought to it from different parts of the farm.

Coffee, ginger-beer, and other non-intoxicating drinks are, I understand, tried in some places as well as, or instead of beer, with advantage, and I feel sure that those who try the experiment of discontinuing the large and demoralizing allowance of beer which has been customary hitherto, will have reason to be thankful that they have done so.

Believe me, dear Sir, yours very truly,

ALEX. W. COBHAM.

Mr. J. Abbey.

FROM SIR PHILIP ROSE, BART., EX-HIGH SHERIFF OF BUCKS.

RAYNERS, PENN, BUCKS.
June 23rd, 1880.

* Dear Sir,—I have much pleasure in replying to your enquiry as to the use of tea instead of beer upon my farm. It was, I think, about six years ago that I unexpectedly went down to the farm in the midst of hay time. An accident had just happened by a man falling off the rick, and I observed the men stupid and sullen, the boys loud and noisy, rough with the horses and pulling them about the head, and the women excited. This led me to enquire into the allowance of beer, and I was appalled at the quantity which it appeared the women and boys were in the habit of drinking. A moment's reflection determined me that if this system was necessary in getting up the hay, I would immediately give up farming; but I afterwards decided upon adopting the new system, and from that time to this I have never given away a single pint of beer on my farm; but as I fully recognized the necessity of some liquid during the exhausting work of hay time and harvest, I have ever since at my own expense, and without making any deduction on account of it, supplied my labourers with cold tea *ad libitum*. I generally purchase a chest of good tea at the beginning of the season, which my bailiff's wife has instructions to make in a copper the day previous to its use: I believe that often a sufficient amount is made for two days. This is put into cans which I procured for the purpose, mixed with milk and sugar, and is sent out between ten and eleven o'clock, and then supplied as required during the day.

The first year the change of system was so successful, that I believe little or no beer was sent for by the men, not even at their dinner hour; but it was a novelty at that time, and I received the thanks of the wives at the conclusion of the harvest. They described its effects as having enabled them to save a great deal more money, and that they had been able to get their husbands up without difficulty in the morning. I have since pursued the system with varied success. One year I found it altogether abused: but this was owing to my bailiff having given in to the suggestion of supplying hot tea for breakfast and at other meals, which was exactly contrary to my object; but I am able to state that the introduction of cold tea has practically brought the consumption of beer on my farm during working hours to a minimum, and that I am quite satisfied that the labourers do more work by the use of tea than by giving beer. It may be that for a spurt, to accomplish an object within a given space of time, a glass of beer judiciously given at certain times may effect the purpose, though I never attempt it: but, as a rule, I am satisfied that the men are in better condition at the conclusion of the day, less stupid and sullen, and certainly much better fitted the next morning to resume their labours, than with the old system of beer. Of course much depends on the bailiff working loyally with his employer in carrying out the system. If he is fond of beer, and especially if he ever commits himself by drinking beer with his men, I should have little hope of effecting much good. Again, it is important that the tea should be good, with plenty of milk and sugar, and then sent out from a cold cellar as required, and nothing can be more refreshing.

I calculate that the cost to me is about £5 a year. Of course, if I were to make any deduction on this account, the system would fail, but an allowance in lieu of beer being added to the wages, the men are enabled to save this if they please.—I am, dear Sir, yours truly,

PHILIP ROSE.

Mr. JOHN ABBEY.

FROM G. G. DIXON, ESQ.,
SWYNCOMBE. HENLEY-ON-THAMES.
June 28th 1860.

* DEAR SIR,—I am truly sorry to have been so long in answering yours of the 16th ult. I am so over head and ears in farming 1,200 acres, I find but little time for anything like correspondence. Yet I am very pleased to give you a few facts. I have now a gang of six men mowing, four of them are abstainers from alcoholic drinks. It is scarcely necessary to say that men can do hard work, not only as well, but really better without intoxicating drinks than with them. I pay them per acre according to the worth of the job. They earned last week £1 13s. 9d. each. I give the abstainers 2s. 6d. extra when the mowing is done. We pay at the rate of 1s. per day, from 6 a.m. to 7 p.m., extra as wages, in lieu of beer. And when we work later, say till 10 or 11 o'clock, we pay 4d. per hour. We always recognize extra work on jobs by money payment and not by giving beer. I make no arrangements for supplying drink. I find, as a rule, that the men like to have the money instead. In the tea and coffee system, some like sugar, others object; some prefer milk, others do not. It is one of my arrangements that no man leaves his work to go for public-house beer. Most of the cottagers brew (without a license) a very simple yet refreshing drink, from ginger and sugar, flavoured with hops or lemon, as preferred, and call it their home-brewed, I wish I could see you to explain any other matters, as I have had many years' experience in abstinence principles, and have seen most encouraging results, alike beneficial both to employers and labourers.
Yours faithfully,

Mr. JOHN ABBEY.   G. G. DIXON.
Steward to Colonel Ruck-Keene.

FROM REV. P. W. LEE.
LEAFIELD VICARAGE, WITNEY, OXON.
July 2nd, 1880.

* DEAR SIR,— I myself have for some three or four years given the men and women (about a dozen) who gather in my small hay harvest, oatmeal drink prepared as you recommend. They like it, and say it is both meat and drink. I also give them 6d. a day in lieu of beer at the end of the whole work. I am sending round about fifty leaflets, &c., with the *Parish Magazine* this month.
Yours faithfully,

Mr. J. ABBEY   P. W. LEE.

CHEDDINGTON, BUCKS.

* Mr J. F. Archer, Parish Churchwarden and Farmer, joined the Cheddington Branch of the Church of England Temperance Society, as a Total Abstainer on its formation in March last (1876). Several of his men and boys followed his example. In harvest-time he paid the usual wages, and gave, in addition, the customary beer-money to the men, 4s. 6d. weekly; and to the boys in proportion. Besides the wages and beer-money, he provided them *gratis* with a plentiful and regular supply of the beverage, the ingredients and preparation of which are stated below. Mr. Archer, who is deservedly a popular man amongst his labourers, was applied to by other men and boys, who offered to work for him in harvest on these terms. Further, the greater number of these, though not Total Abstainers, but regular beer-drinkers declared their willingness not only to abstain during their harvest work, but also, without joining the Temperance Society, to forego their old habits of beer-drinking as long as those who had joined did so. This voluntary

abstinence for the most part, and in one or two specially uncertain cases continues. The result of all this has been a decided success. Not only were the men enabled to work heartily and well together through the entire harvest, without bickerings, disputings, or any unpleasantness, but they did so, as they willingly admit, without in the least degree wanting beer, and as being better satisfied, better in health, able to do more work, less fatigued, and, moreover, considerably better in pocket, than with it. The wives of the men and the mothers of the boys are unanimous in declaring that they never had so pleasant and profitable a harvest-time; and they are grateful for the working of the Temperance Society in general, and for Mr. Archer's excellent development of it in particular.

I cannot close this paper without adding, that the men distinctly state that they could not have borne the heat and fatigue of harvest work without beer as before, if Mr. Archer had not provided them with a beverage in its place; in other words, if left to their own resources, they could not, however desirous of doing so, have found any other drink than beer, and therefore, for this simple reason, beer they must have had.

As a consequence, plainly enough deducible, I would point out that our Society must address itself as forcibly and as influentially as it can to the employers of agricultural labourers, to enlist their sympathy with, and to lend a helping hand to them. The men may be very willing, but they can practically effect little unless the masters come forward to lead and assist them.

FREDK. BURN HARVEY,

May, 1877.   RECTOR OF CHEDDINGTON, BUCKS.

BEVERAGE ABOVE REFERRED TO.

To make one gallon : Take half a pound of oatmeal, quarter of a pound of cocoa, and half a pound of sugar ; mix with a little cold water into a thick paste; put the whole into a gallon of boiling water, or into hot water, and boil it.

The men like it best warm, as more pleasing to the palate and more effective in quenching thirst and even satisfying hunger. If used cold, it must be fresh and never more than a few hours old. The cost of this gallon is cheap enough : half a pound of oatmeal, 1½d. ; quarter of a pound of cocoa, 2½d.. ; half a pound of sugar 1½d. Thus 6d. per gallon, or 1½d. per quart, is the outside cost of the beverage. It was given two or three times a day.   F. B. H.

BUSCOT WICK, FARINGDON.

* DEAR SIR,—I have received your letter in which you referred to a conference about to be held in the Corn Exchange, Wallingford, on the 2nd of March. I shall not be able to attend personally, but as I approve of its object, I will very readily give you my experience.

To begin with myself. I have been an Abstainer for thirty-three years, and think that if I abstain from all alcoholic drink myself, I cannot conscientiously offer them to others. Moreover, I am well aware that any kind of farm labour can be done better without beer than with it. I am at the present time farming upwards of 200 acres, and consequently employing a good deal of manual labour. The wages I am giving at the present time are fourteen shillings a week, with a good house and garden rent free, and in the harvest piece-work as long as it lasts. During hay-making and harvest only do I give any perquisites : we then make some tea, with plenty of new milk and sugar ; and when late of an evening, their supper; and I can assure you that I have never yet had one complaint made from any one man of my treatment in not giving beer to those in my employ, and I have men that have been in my employ now for five years. The best testimony I can give will be that of my carter who will have been with me three years the 6th of next April. I will ask him to write a few lines stating his experience. When he first entered my service he was an habitual drunkard. He has been an Abstainer for nearly two years, and is now an honest, hard-working man ; whereas before he was only an injury to himself, his wife and children badly clothed and ted,—all through drink. The change has been beneficial both to themselves and their master.

I do earnestly wish that more of my brother farmers would pay their labourers' wages all in hard cash, doing away with perquisites, especially beer ; for, as I have before said, all my men are satisfied.

More than that, if I have to dismiss a man from regular work, I have no occasion to go to a hiring fair to supply his place. I think that if paying men in cash, giving no beer, was to be the rule and not the exception in farming, we should remove a temptation to drink from the working man, and get better labourers, by keeping the drink from them when it is in our power to do so.

I remain, yours sincerely,
Feb. 23, 1877. J. WALKER.

LETTER REFERRED TO IN THE ABOVE.

DEAR SIR,—My master having desired me to write to you respecting agricultural work being done without beer, I must first inform you that I for many years was a drinker of beer in my work, but for this last two years and upwards I abstained from it altogether ; therefore I can speak from experience, and that is, I am certain all kind of farm work can be done better without beer than with it, and that men would find, by giving it a fair trial, that their work would be done easier without the beer than with it.

I remain, yours obediently,
BUSCOT WICK, Feb. 24, 1877. JOHN JORDON.
Mr. JOHN ABBEY.

Acacias, Reading, June 18, 1881.

DEAR SIR,—I have much pleasure in answering your enquiries. I have had many years' experience and consider that the men do their work better without beer. I pay an equivalent in money, and sometimes give cocoa and tea extra. I have farmed about 700 acres, in two distinct districts, 500 and 200 acres. The plan has answered well for employer and men. It would be well, I think, if there were no drink sufficiently alcholic to stupify men.

Yours truly, GEO. PALMER,
Mr. JOHN ABBEY. (M.P. for Reading.)

ASTON ROWANT PARK, TETSWORTH, OXON, June 14th, 1881.

DEAR SIR,—I have pleasure in answering the questions contained in your note of the 13th inst., and will reply to them in the order in which you have put them.

1. It is 18 years since beer was allowed on this farm of 600 acres.
2. The quality of the work has not deteriorated during this period ; there is, I think, less time wasted than formerly, and the men less quarrelsome than when beer was allowed.
3. No beer money has been given. All harvests, and much of the farm-work, being paid by the pieces, the men supply themselves with beverages, but alcoholic drinks are not allowed on the farm, nor smoking.
4. The men seem satisfied with the arrangement, and nothing would induce me to return to the old system of part payment in beer.
5. The number of hands I employ is, on the average, 35.

I remain yours truly, THOS. TAYLOR, (J.P., Oxon).
To Mr. John Abbey.

KINGSTON BLOUNT, TETSWORTH,
DEAR SIR, 16th June, 1881.

1. I have never given beer or beer-money, I pay so much wages, and the men do as they please.
2. I do not think it makes any difference in the work.
3. I always pay money. I have never tried other kinds of drink.
4. My usual number on the farm is about 20. Many more in harvest.
5. I am sure it is much best for both men and master.

I think, as a rule, the men working together mowing, drink beer. In harvest, working separately they drink tea and coffee I believe if you can sell tea and coffee ready-made in a village, good, and cheaper than beer, the men would buy tea and coffee instead of beer. I shall be pleased to give you any further information, and should only be too glad if all employers of labour would stop beer or beer money. Yours faithfully,
Mr. ABBEY. ARTHUR H. C. BROWN, (J.P., Oxon).

ADWELL HOUSE, TETSWORTH, OXON,
June 13th, 1881.

DEAR SIR,—My abscence from home has delayed your letter. I am afraid that my experience will not aid the object that you have in view. For the last 12 years or more, I have always paid my labourers in hard cash, instead of beer, in the hope that they would expend it in food instead of drink. I find they work well and sometimes save their money. On long and hot days I give them tea in the evening as an extra. I am afraid that they would not leave off beer entirely in the day time. I should like to know what substitute can be given. Yours faithfully,

H. BIRCH REYNARDSON, (J.P, Oxon).

BULSTRODE ESTATE, GERRARD'S CROSS,
July 23rd, 1878.

DEAR SIR,—I thank you for sending me the new leaflet about tea in the harvest field. The circulation must do good. I do my best to circulate the *British Workman* and other papers of a temperance character amongst the workmen in this locality. *This year on this estate we have paid the men in the hay-field money instead of giving them beer.* Through the kindness of J. Bramley Moore Esq., we have a very excellent free reading room here, for the working men. Yours truly, HENRY WITHAM.

NUNEHAM COURTENAY, OXFORD,
June 16th, 1881.

SIR,—In answer to your letter of the 13th instant, I beg to send you the following answers to your questions :—

1. We have not given beer for ten years.

2. During the hay season cash is given instead of beer ; but all the harvest work is done by piece-work, when neither beer-money nor beer is given.

3. The work is more satisfactorily carried on, on the above system, than when beer was given. In some instances the old plan of giving beer worked very badly, to say nothing of the men taking too much ; for some of the men did not care to have beer, so that they had neither beer nor money ; but now those who like to save their money are able to do so, and those who will persist in spending the money allowed for beer, do so on their own responsibility at any rate.

4. I consider the system, as explained above, has worked very satisfactorily.

I have, during our hay-making, which season was terribly hot, had tea sent to the men every afternoon, in addition to the money payment, and I found that the plan was thoroughly appreciated.

I am more in favour of giving an allowance in money, than in any other plan ; it will not stop the older hands from having beer, but it gives the younger ones an opportunity of not contracting a habit of drinking on every occasion. Payment in cash gives the young labourer a chance of becoming a depositor in the savings bank, and from a " thrift " point of view, is very much more useful than giving tea, cocoa, coffee, &c. Another thing, it is a much more business like arrangement between employer and employed ; the employer knows what he gives, and the employed what he will get.

Faithfully yours, F. MAIR,
Mr. J. ABBEY, (Steward to E. W. Harcourt, Esq., M.P. for Oxon.)

CHALK PIT FARM, ENGLEFIELD, READING
June 11, 1881.

DEAR SIR,—I will with pleasure answer your questions.

1. How many years have you abandoned the beer ? Six.

2. If you consider the men do their work as well ? Quite so ; I may say better.

3. Have you paid in hard cash, or have you paid part in money and given some other drink as well. If so, what sort of drink did you give, and how did you manage it ? We pay in cash, 1½d. an hour extra in haytime and harvest, or 1s. 6d. per day if we make long days ; from seven till six, 1s. per day. It is my opinion labourers

ought to have what they earn in cash, the employer has no right to compel his men to take what he thinks is proper, the workmen know what suits them best, and they make drink according to their palate. Some of ours have milk and water, others have made ginger beer, tea, &c. There is very little drinking going on at work. They begin to find that the less they drink of anything the easier they can work. I helped to stack all the corn and hay grown here last year, and all the drink that was brought on to the farm was tea at five o'clock; they also have a little skim milk at that time. All our skim milk is given to the labourers (something like twelve gallons per day), they fetch it daily, some in the morning at seven o'clock and the others in the afternoon at five o'clock. We commence work at seven, the men leave at twelve for dinner (one hour) their homes being near or on the farm; commence again at one, and leave at half-past five. If we are busy at hay or harvest work we take tea at five and keep on till eight or after. The average pay for overtime, including beer money, is 1s,6d. per day, and some of our people after harvest is over buy themselves new clothes or articles of furniture. Several have said to me " We should not have had this or that if we had been obliged to have had the beer." I try and select active civil men and avoid strangers and loose characters.

4. Has it answered well for yourself and men? Admirably. You may depend I shall never go back to the old system. I know too well the trouble connected with the beer system ; it keeps one employed to draw the beer, wash the vessels, and see that it is served out properly, as well as a constant grumbling, the quality not suiting the men, although when I gave beer I had mine made in the right season and I gave it them in good condition, still it was not satisfactory. New or sour beer must be very injurious, good food and pure water I find the best thing to work on, and I have done as hard work as most people. From between when I was twenty to twenty two I took nothing but water to drink and I was very strong, and now I am past fifty very few are so strong. I am not a strict teetotaller, but the less I take of intoxicants the better I am. •All sensible men know that they do not derive any benefit from alcoholic drinks ; the reason they take it is because they like the flavour of this wine or that liquor. I will just mention that the late Mr. J, Badcock, of Thame, a very successful farmer and good manager, occupying more land than any other farmer in the parish, gave no beer but paid his men 1s. per day in lieu of beer. All the other farmers gave beer, but Mr. Badcock had the best staff of labourers and got over much more work than his neighbours. That was from 1844 to 1852. The reason he adopted that plan was when he gave beer the men were constantly quarrelling, and I think one man was killed from fighting through the beer.

I have managed Mr. Benyon's Home Farm (750 acres) nearly six years, employing about twenty-five hands, and have given no beer the whole of the time. We have about 400 acres of hay and corn to harvest, and my experience is this, that it is more satisfactory to the labourers and much less trouble to the employer. We are not annoyed with the complaint of spilt beer or thick and sour beer, for without very great care in keeping the jars and cups constantly washed in hot weather the beer becomes unpalatable and unwholesome.

I farmed on my own account 18 years at Shinfield. All that time I gave beer through the hay time and harvest. My custom was to have the beer brewed in February and not tapped till early in June. I had it made by one of the best Reading brewers, the cost was 1s.4d. per gallon. A light pale ale, clean drinking, and always bright and sweet. I was very particular in having what we consider pure beer, I also attended as much as I could to the drawing of the beer and serving it out, still we had unpleasantness at times. I am employing a man now that was with me 16 years at Shinfield, a good sober labourer, and I have known him take when he has been pitching corn all day 10 or 12 pints of beer. This same man pitched nearly all the hay and corn at Chalk Pit, and I never saw him take beer the whole of the time. A man that has been working here 21 years said at the finish last harvest that the work was done more comfortable and pleasant than he had ever known it done before, without any beer at all coming into the place. That man is not a teetotaller, but a steady, respectable, moderate man.

I have had as much experience as most people, and I am certain work can be done better without beer than with it. If people think they ought to have it, it would be better for them to take it after the work is over.

I am, yours faithfully,
THOS. E. SHRIMPTON.

Mr. Abbey.   Manager to R. Benyon, Esq., ex M.P. for Berks.

BIRTLES HALL, CREWE,
Dear Sir,   July 23rd, 1881.

1. If the men are properly fed, I feel confident they will do as *much work* and *as well* without beer as with it.

2. I would advise a cash payment in all cases where practicable. But if the moral standard of the men be low a substitute might be better for both master and men. Where the men are boarded in the farm house I should advise a substitute,

3. Not to give stimulants would be an incalculable benefit to working people and more satisfactory to their employers.

4. I have farmed in three counties, East Riding of Yorkshire, Cheshire and Shropshire since 1836, on farms varying from 800 acres to 250 acres, besides employing navvies in reclaiming land; also bricklayers, masons, brick and drain-pipe making, woodmen, &c., and have employed men from many counties in England, Wales, Scotland and Ireland, and have invariably found as a rule, that the men who cared the least for beer or cider were the best and most trust-worthy workmen.

5. At present I am not farming on my own account, but for a gentleman for whom I am agent. We only employ from 10 to 12 men, and they have asked to have money in lieu of beer in the ensuing hay harvest.

I have long since came to the conclusion that no person in health needs a stimulant, and to the young alcoholic drinks are pernicious, both morally and physically. When once a youth of the working classes takes to *beer and tobacco*, I predict that he never gets upon the first step of the social ladder.   I am Sir, yours very obediently,
Mr. J. Abbey.   L. WHITTY.

KENDRICK HOUSE, READING, 24th June, 1881.
DEAR SIR,—I have pleasure in sending you the following answers to your enquiries which I hope will be of use to you.

1. I have farmed five years without giving beer.
2. I consider the men do as much work, and as well.
3. I have paid in cash. As a Temperance Reformer, have given tea and other drinks to convince the work people by actual practice that beer is not necessary.
4. The plan has answered well for both employer and men.
5 Number of hands employed, 8 to 12. No difficulty in getting labourers to work without beer, no trouble has ever occurred in consequence of no beer being served out.
I am, dear sir, yours truly,
WM. PALMER, (J.P.)

FAIRSPEAT HOUSE, WITNEY, OXFORD, July 14th, 1880.
DEAR SIR,—I have received your letter and enclosures. No doubt my plan of *Money Payment and no Beer*, for various work during the year performed by manual labour, tends to sobriety.
I find no drink whatever. They consume what beverage they choose. All money payment is by far the best, both for masters, men, and their families. I have never had any difficulty with the men. The farm is 1,300 acres.   I remain, yours respectfully,
Mr. J. Abbey   JOHN S. CALVERT.

' POTTERS HILL, NEAR WITNEY, July 30th, 1881.

DEAR SIR,—On coming into this neighbourhood about 15 years ago, my father and I discontinued the practice of giving drink in part payment for work during the hay and corn harvest, and on the three farms with which we have to do, employing from 30 to 60 persons, we have found the plan work well, nor would we or our men on any account go back to the old system. The time and labour taken up in drawing and distributing the beer, over a distance perhaps of 2 miles, is in itself a consideration, when every hand and hour are precious. If the men have money instead, each provides his own drink, and brings it in the morning with his food and tools. Many men when beer is given feel that they have not had their full days' pay, if they have not drunk their full quantity of beer. The drink-giving system is manifestly unjust towards those who cannot take beer, while total abstainers naturally feel that they are hardly done by under it, and I believe this has been the cause of many breaking their pledge. A neighbour of mine informs me that a boy he employed, who is by no means what would generally be termed a " bad " boy, got at the beer jar on the sly in the harvest field, and was drunk and incapable for the remainder of the day. If farmers would pay for all labour in cash, they would offer a less convenient handle for union and other agitators to lay hold of; these as a rule conveniently forget payments in kind.

Yours truly,

J. ABBEY.   BENJAMIN HOBBIS.

A SALE OF SHORTHORNS WITHOUT GROG.

At Epsom, on the 19th of May, a sale of valuable " shorthorns " took place from the farm of a good temperance man, Mr. E. C. Tisdall, of Holland Park Farm and Epsom. The sale was attended by buyers from various districts, one at least being from Ireland, and the value of the stock was estimated at two thousand and eleven guineas. At the luncheon preceding the sale no intoxicating liquors were supplied, and the absence of these customary articles did not seem to be regretted by the visitors. Mr. W. S. Caine, M.P., who filled the chair, pleasantly reminded the gentlemen that nothing was provided which could disturb their perception of the true character of the stock; and at his suggestion—though other beverages were liberally furnished—the toast of the Queen was drunk with all the honours in " honest water."   Mr. Tisdall's example is one which all owners of stock ought to follow, and all purchasers ought to encourage.—*Alliance News.*

HARVEST HOME WITHOUT BEER.

May 1st, 1882.

DEAR SIR,—Our Vicar has brought us your Pamphlet, " Intemperance." We think it calculated to do much good, and would like 50 copies to circulate amongst our own men and neighbouring farmers, many of whom still think the Beer must be given if the work is to be done. We have for several years given tea and oatmeal drink to our harvesters, and paid them money. When the seasons are good, we always have a Harvest Supper—from 30 or 40 (men and wives), all who have helped to turn a sheaf. Plenty of roast beef; plum pudding *ad. lib.*, with tea and coffee in abundance. The men make merry without doubt, and, to finish, my boys and girls take the piano in and dress up the barn and sing, and the men and women sing too; ah! much better than they would with hogsheads of beer, as of old. Still much has to be done in every village and town, and your book will help to convince many. One of my boys has a Band of Hope of 120, another is Secretary of the Church of England Temperance Society just formed, but we want hundreds and thousands of helpers, looking at the state of the country generally.

Yours truly,   A. S.

VALUABLE TESTIMONY.

THE EARL OF JERSEY said, with regard to the use of beer in the hay and harvest fields, he might say that for four years they had discontinued the practice of giving men beer during the harvest months on his farm. The men were given money instead. His agent. Mr. Little, knew that the men liked this arrangement, which was advantageous in more ways than one. There were now no disputes as to the quantity of beer each person should receive, and no time was lost in fetching the beer. Indeed he thought it would be difficult to carry on a large farm like his under any other system. He was bound to say, in justice to the labourers, that he had a most excellent body of men. He believed that every employer of labour who tried this plan would be satisfied with the system ; indeed he believed he was not the only person in the neighbourhood who had adopted this custom, and found it to be a satisfactory one.—*Speech at Bicester Conference, December 19, 1881.*

THE HON. PERCY BARRINGTON said it was a common impression amongst labouring men that they could not do hard work without drink, but he was happy to say that was an exploded idea. He thought it had been proved that they could work equally well without it. He had done all he could in that direction himself ; he had paid his men money instead of beer. He supplied no beer, but in its place he gave them as much tea as they liked to drink gratis. He had found the greatest benefit from it.—*Speech at Buckingham Conference, October 31, 1881.*

SIR HENRY DASHWOOD, BART. said he had been farming for a good many years, and he had not given beer to his employés for a long time. His boys and men found their own substitute. He should like beer kept off the ground altogether. He had found that beer stopped work. He had also asked a good many farmers about it, and had been rather surprised to find sometimes that men whom he had thought would have given beer were against it, because they found, as he had found, that, instead of stimulating, it stopped work ; and that was Dr. Andrew Clarke's opinion about it. He says that however pleasing alcohol may be to the taste it is not a helper of work. On the contrary, it is not only not a helper of work, but it is a great hinderer of work. He now always gave money to his labourers instead of beer. It was best for the labourer and for the employer also. The men did their work better without the drink, and, besides, when the money was paid they were not tempted by their employers to drink. He thought that was one thing which they should all be very careful to avoid—the placing of temptations in the way of their men.—*Speech at Chipping Norton Conference, January 23, 1882.*

W. M. FOSTER-MELLIAR. ESQ., J.P., said he did not believe that alcohol helped a man in his work in the harvest field. It stimulated him, he admitted, but it afterwards left him more exhausted than what it found him. In speaking of this, he might use a homely but a very true analogy, and that was the poking of the fire. By poking a smouldering fire they produced heat, just as alcohol acted upon the body, but unless they added fuel as well as poked the fire, their fire would very soon go out. This was precisely the case with alcohol. It created temporary warmth and brightness, but it added no new coal to the flame. Speaking of the labours of the men in the field, he said he could speak from experience that beer could be done without. He had often had a hard day's work shooting and walking long distances, and he had proved beyond a doubt that he could perform this task with far greater ease on cold tea than he could on alcohol. He left it to others to say what the drink that should be supplied to the labourers should be, but he maintained that the supplying of beer should be no longer continued. They might give the men the money, and let them decide for themselves what they should drink.—*Speech at Banbury Conference, November 22, 1881.*

Dr. W. Collier, as a medical man, addressed the meeting on the same subject, and said he objected to giving men beer in the harvest field, because it injured the labourer, injured his work, and injured his pocket. As a worker in a large provincial hospital, he had seen the diseases produced by intemperance, and in many cases he had traced it to the drinking in the harvest fields. He advocated the substitution of tea or oatmeal fluid for beer, and said that a good cottage and garden would act as powerful counter attractions to the public house.—*Speech at Banbury Conference, Nov.* 22, 1881.

Mr. B. Hobbis, as a practical farmer, said he was not going to say intemperance was the cause of all the agricultural distress, but he believed it was an important element in their many troubles. He did not believe the farmers would suffer very much by the discontinuance of barley for malting purposes, but if the current of drunkenness was stopped, they would be very great gainers by the reduction of local taxation. He condemned what he called the senseless practice of treating over bargains. He urged upon his brother farmers to discontinue the practice of giving beer or cider to their men in the harvest field, because it was to a great extent, the cause of much drunkenness among their rural population.—*Speech at Faringdon Conference, December* 12, 1881.

Mr. A. Blake (estate steward to A. Brassey, Esq., of Heythrop Park), said on the estate of A. Brassey, Esq., he had for the last seven years given the men extra money in the place of beer, and he found it work better in every respect. The men were better satisfied, they did more work, and did that work better than when they took the stimulants to them in the field. He found that the men preferred to have the money instead of the beer, and the amount which he allowed them in lieu of beer was tenpence a-day. He said that threshing was a dusty job, and he allowed 4d. a.day to each man. He had great pleasure in seconding the resolution, "that this conference considers that it is desirable that labourers working in the hay and harvest field should be paid in money instead of beer." The resolution was carried unanimously.— *Speech at Chipping Norton Conference, Jan.* 25, 1882.

Bearwood Farm, Wokingham,

30th January, 1882.

Dear Sir,—I have farmed for several seasons past about 2,500 acres of land for Mr. John Walter, M.P., and others, and my system is money and no beer, not merely in hay and harvest time, but all through the year, for work for which beer was formerly allowed. I do not provide any tea or other drinks as a substitute, thinking it better (in so large a business at any rate) to treat the matter on purely commercial grounds, giving increased pay for extra work rendered, and leaving the men to use their own judgment as to the nature and quantity of their drink, the same way as they do as to food. If I attempted to alter the system from money back to beer, as formerly, my labourers would object to a man. Under the money plan our work is done better, cheaper, and far pleasanter; many of the men are abstainers, and those who are not drink only with their meals, as a rule. The tenant farmers are adopting the money system, and I have never known one who did wish to change. They only wonder how on earth they could have kept up the beer system so long. The first season always suffices to thoroughly recommend the plan of money in lieu of beer to both master and man. Our largest tenant on the estate farms in all 1,400 acres of arable land, and, having followed my plan, now some years, would endorse every word I have stated in favour of it.

I beg to remain,

Yours very faithfully,

H. SIMMONS.

Mr. John Abbey.   Steward to John Walter, Esq., M.P.

Park Place, Henley-on-Thames, Jan. 31, 1882.

Dear Sir.—In reply to your enquiry of the 28th inst., I may tell you we have discontinued giving beer during harvest for some years, and, instead, the men get tea and oatmeal drink, and 1s. "beer money," as it is called; but I have given directions that, in future, instead of 1s. the men shall receive 9d. only, thinking the drink is quite equivalent to the difference; and this, I trust, will be an inducement for farmers and others to follow my example, which they would not do if I gave the 1s.—I am, faithfully yours,

Mr. John Abbey. JOHN NOBLE.

Captain TROTTER, of Dyrham Park, near Barnet, says :—"Having tried the system of total abstinence last year during the hay season with my men. I have found it answer in every respect perfectly. I had my men from Bedfordshire, and, having calculated the expense of the former allowance of beer per day per man. I gave them exactly the same amount in money, and my bailiff assures me that nothing could be more regular than the men, and on Monday morning, instead of being weaker, as formerly, from the effects of Saturday's and Sunday's drinking they were refreshed and stronger than ever ; that he never had an angry word during the whole season, and never heard an oath ; and such was the success, that I shall never have any more beer in my fields, and I know that I shall be as much benefited by the steadiness of the men as the men will by the saving of their constitutions and money. Many persons came during the hay season to see the dinners go into the hay-fields, which one of the men cooked at their expense for his time, and instead of cans of beer and a little bread and cheese, a large wheelbarrow full of roast and boiled meat, in large pans, and potatoes, &c., and a pailful of coffee were sent to them. Two or three of my neighbours tried the same plan with similar success."—*Aylesbury News.*

FARMERS AND TEMPERANCE.

A meeting of a novel character was held at the Temperance Hall. Frome, December, 1877. All the addresses were given by farmers. Mr. D. B. JOYCE, of Beckington, presided, and referred to the change which had been wrought in public opinion. He then adverted to the immense waste and destruction of grain in the manufacture of alcoholic drinks, which were admitted by the leading men in the medical profession to be almost useless. For considerably more than thirty years he had not given alcoholic drinks to his men, and he had been an abstainer himself about forty years. At first he met with a good deal of opposition, but it was soon found that any kind of farm work could be done on temperance principles. After a hard day's work, teetotal labourers were better fitted for work the next morning than those who took stimulants.

Mr. G. JARVIS, of Kilmington, said he could boast of twenty-four years experience, and on looking back upon his past life he felt ashamed that he did not adopt teetotalism earlier, although he never liked alcoholic beverages. He signed the pledge, for the sake of example to his children and others. He told his men he should not give them any more drink, but he would give them its equivalent in money. The small beer which many farmers gave their men cost about 6d. a week for each in the winter ; he gave his men 1s. extra in the winter and 3s. in the summer. He only had one man leave him in consequence, and never found any difficulty in carrying out his plan. He would like to see the men who would not work as much for money as for beer. When the harvest time came he once found that his men had sent for a jar of cider, but he told them he should not allow that again, and that if either of the men could not work without cider, he would pay him his wages. One of his men saved his beer money and bought a load of coal, while his wife saved hers and purchased him a pair of trousers. About eight years ago he took another farm a long distance from Kilmington, where there were eleven acres of orchard and a good crop of apples. When the apples were ripe people asked him what he was going to do with them, for of course the rent had to be paid. He told them he should make them into butter and cheese. The season was a very bad one for haymaking, so he set the chaff-cutter to work cutting up straw, ground up the apples and mixed the two together. Nothing could answer better, and his butter was pronounced to be as good that winter as in the summer. The speaker then narrated some of his experiences as a poor law guardian for twenty years, during which he had had many a battle on the question. In some cases where sick paupers had been prescribed alcoholic drinks without any beneficial effect, he recommended mutton instead, and the men were soon able to resume their work.

Mr. Louis Vallis, of Hemington, said he was thankful he had been a teetotaller for four years. During certain seasons of the year, farm work was very hard and the days were long. He had heard some people say—" It is all very well for town people to be teetotallers, but let them come into the fields and pitch hay all day when the sun is hot." Well, he could pitch hay all day without strong drink, and feel well and hearty after it. He believed he could do as much work, take the year through, as any farmer. He generally carried with him into the field a bottle of water or coffee, and he could feel quite refreshed by it. In the winter time he never wanted to drink between meals.

Mr. Watkin, bailiff at Dilton Farm, Westbury, said he could reap, plough, sow, or mow without strong drink, and he had done so for nearly thirteen years. He could safely say he was better without the drink than with it. He had pitched nearly eighty or ninety sacks of wheat a day without a drop of alcohol, and as his men knew he was a teetotaller they determined to test him to the utmost. His men sometimes told him if they could live as well as he they could work as well; but he told them they might live as well if they expended their money in beef instead of beer. He had sown sixty-four acres of land per week, often carrying about 10lbs. of dirt on his boots. He could do anything in the way of farm work without alcohol, and thoroughly enjoyed life. He generally drank cocoa as a beverage, and he was certain that a quart of cocoa would do a man much more good than a quart of beer.

Mr. Hampton, of Potterne, near Devizes, said thirty four years ago last month he became a teetotaller. The farm he first occupied was a dairy farm, and he milked over seventy cows. His men asked him what he should do in the haymaking time. He told them he should not give them any beer, but the money instead, and he believed they made more hay than any other farmer in the village. All the farmers in the neighbourhood were opposed to him at that time, and their men annoyed his men. Two or three young men were very abusive to him, but he offered to forfeit £10 to their £1 if he and his six men did not do more work in a given time on teetotal principles than either of them and six other men would do on alcoholic drinks. They would not accept his offer, and he was not annoyed afterwards. For twenty-nine years he had not heard an oath or any discontent on his farm. The greater part of his men became thorough teetotalers. The second year one of his men had a fat pig to kill. and he said he used to have to sell it to pay his rent, but now he was going to keep it for his own dinners. He afterwards had a sheep and corn farm of between 700 and 800 acres. One of his sons, who had never drank a drop of alcoholic liquor in his life, weighed eleven score. He had just taken on the management of his farm. Another son, also a life teetotaller, had taken a farm of 100 acres. There were not two stronger young men in Wiltshire.

Mr. W. Dew, of Beckington, gave an account of his adopting teetotalism twenty-seven years ago. He used to say that all teetotallers ought to be drowned, and people told him he would soon be buried if he did not give up drinking. He suffered from indigestion and an overflow of blood to the head. A person recommended him to become a teetotaller, and he promised to give it a fair trial. He went to the doctor and told him so, and the medical man told him he would not want to see him again for two years. He had never had occasion to go to him since. The chairman gave him a work entitled "Anti-Bacchus" to read, and he was thankful for the light he obtained from it. He had uphill work amongst his men for some time, and some of them left him. He gave them money instead of beer, and they soon came to their senses. He had sown eleven acres per day and twenty sacks of corn. There was no farm work that could not be done without strong drink. He believed the drinking customs were the cause of the present depression in trade,

for if the money spent in liquor were expended in useful articles there would
be a wonderful revival in trade. A couple of men could soon brew enough
beer to ruin thousands.

Mr. CLARKE, of Buckland Dinham, said he had been a teetotaller for
thirteen years. When he signed the pledge, his young master told him he
ought to be shot or hung, but his old master said if he heard such an
expression again he would "stop the tap" altogether. The speaker then
enlarged on the advantages of teetotalism in all kinds of farm work.

WHAT THE FARM LABOURERS SAY.

* DEAR SIR,—I am greatly pleased with what you are doing to do away
with paying men engaged on the land with beer. I can bear my testimony to
the harm that has been done to the men, and especially the lads. I believe
that thousands get drunk for the first time in the hay and harvest field, on the
drink given by the master. Vast numbers learn to like the drink in this way,
and so all that they have learned at school and church is undone, and a life of
sin and poverty is commenced. The practice of giving men, women and boys
beer as has been done in the past is wrong; it is unjust; it is cruel and
demoralizing. The idea is that more work is got out of the men, but it is not
so. I can speak from experience, and I know that a man can do more work
without the beer, because he can then get proper food with the money saved
from beer. Every working man knows that it is food that gives strength.
The beer only takes the strength out of him; then more than this, men get
drunk in the field, as Sir Philip Rose said in his letter, and if they have a
public to pass on their way home, they cannot get by, they often go on drinking till they are turned out, and go home drunk at a very late hour. What
state can such men be in for their work the next day? Then the beer system
often leads to quarrelling and fighting. I remember once two men, after they
had had their beer at four o'clock, quarrelled and fought with their scythes,
and one cut the other fearfully, and I assisted in sowing up the wound in the
poor drunkard's body. I have had a good deal of experience in farm work,
and I am satisfied that you are doing a work that will prove to be a real
blessing, both to employers and labourers, and if the society had been at work,
as it is now, 20 or 30 years ago, the labourers would have been in a very much
more comfortable and contented position than they are at present. I cannot think
that thoughtful employers will go on in forcing the beer upon the labourers,
when they know what a curse it is to men and their families, to both body
and soul. The church cannot do a better thing. God is blessing your work,
and I feel sure that He will bless it.

Your humble Servant,
A WORKING MAN.

MOWING THREE ACRES OF CORN IN ONE DAY, WITHOUT BEER.

It will be remembered by our readers, that at a Conference, held in the
Music Hall, on May 29th, in connection with the Leamington branch of the
Church Temperance Society, and under the presidency of A Hodgson, Esq,
President of the Warwickshire Chamber of Agriculture, that Mr. J. Abbey,
of Oxford, in a speech stated that he had known a man mow 3 acres of corn
in one day. This statement was challenged by Mr. Wakefield, of Fletchamp-
stead. Mr. Abbey accepted the challenge, and stated that he would give £5
to the Leamington Hospital if he failed to establish his statement, if Mr.
Wakefield would agree to do the same if he succeeded. It appears that
Mr. Abbey sent the report of the conference to several people in Yorkshire,
and the following letters have been received, which the vicar of Leamington
has forwarded to us for publication. It will be seen that Mr. Abbey's statement is fully sustained, and to prove that the letters are *bona fide*, names,
and addresses are given.—*Leamington Chronicle.*

My dear Sir,— In answer to your enquiry, I am glad to say that I have
frequently mown 3 acres of corn per day. I, and my brother George, once
cut 7 acres between us in one day, and at the time were honest teetotallers of

20 years standing. Our chief drink was oatmeal and water. I may also state that I have known other men on the Yorkshire Wolds cut 3 acres of corn in one day often. I have mown 2 acres of grass in one day for weeks together. Working along with other men who drank beer, I cannot remember except in one or two cases where they could work with me. As a rule they had to give up in the middle of the day and get under the hedge, especially if the day was hot. I think this confirms your statement, and I am not giving something I dreamed, but real facts.

I am glad you have such a good job in hand, and that your Yorkshire pluck lives. Many thanks for your kind enquiry, through the mercy of God I am doing well on the whole.

Yours very truly,
WILLIAM ROBINSON, (Farmer),

(Mr. Robinson is now farming for himself). Brakes Farm Huntingdon, York.

June 7th, 1879.

P.S.—The grass mowing was done for Lord Herries, Everingham Park.

DEAR SIR,—I am glad to find that you are so strong in your belief. It is a fact that three of us have mown 3 acres of corn per day each, for days together—George and William Robinson, and John Atkinson, without any of the drink. We have never taken any of any kind for 38 years, and we love the temperance cause to-day as well as ever. I hate the drinking system, every inch of it, and I am glad, sir, to find you so strong at the present. Go on, Mr. Abbey, with your big guns, and down with the drink traffic. If I can help you at any time I will do it with pleasure. Please write and tell me how you get on.

Yours truly,
GEORGE ROBINSON,

June 7th, 1879. Le-street, Thorpe, Market Wrighton, Yorkshire.

DRINK BEATEN IN THE BRICK-FIELD.

MR. ABBEY, SLOUGH, June 11, 1875.

DEAR SIR,—I enclose you a little information, which speaks for itself.

We have 11 abstainers in our field, and the result will be something worthy of our cause. The summer is the trying time for Brick-field men, and if our work here does nothing else it has supplied me with an argument so practical as to shame those who say they " Can't work hard without strong drink." 1,000 burnt bricks average a weight of 2 tons 8 cwt., and a man in making that quantity must lift a heavy mould that number of times, and lift the green bricks twice, or nearly 6 tons per thousand, and some of our Temperance men in a fine week make 60 thousand. Of course this is not the average. This is labour without strong drink ; not a man could do it with the drink. A moulder, who is not a drinker, invariably can keep his gang together.

Work done, week ending June 5, 1875 :—

By Abstainers.		By Drinkers.	
No. of Bricks made.		No. of Bricks made.	
No. 1 Stool	 420,000	No. 3 Stool	 311,500
4 ,,	 384,000	*5 ,,	 374,000
7 ,,	 335,000	6 ,,	 298,000
†2 ,,	 330,000	9 ,,	 301,000
	1,469,000		1,284,500

N.B.—I have picked out the 4 best on each side.

* This man is not a pledged Abstainer, but he has not had any drink since we began brickmaking. viz., 2nd week in April.

† This is a steady man, but his quantity is small because his gang broke up two or three times.

Yours truly,
H. J. DANS.

C

ACCIDENTS AND DRINK.
(*Extract from a letter*).

A sad case of death from drunkenness occurred here on Friday, a waggoner being run over and killed by his waggon, leaving a widow and six children. This is, I think, the third death that has occurred in this parish from a similar cause, though neither of the men were parishioners. Wishing you every success in your good work, a work which God will bless,

I remain, yours very faithfully, E.P.

DRINK OUT-DONE IN NAVVY WORK.

SIR THOMAS BRASSEY, M.P., in his book on "Work and Wages," says: "The taste for drinking among a large number of working people in this country has been excused on the ground that hard work renders a considerable consumption of beer almost a necessity. But some of the most powerful among the navvies have been teetotallers. On the Great Northern Railway there was a celebrated gang of navvies, who did more work in a day than any other gang on the line, and always left off an hour earlier than any other men. Every navvy in this powerful gang was a teetotaller."

THE VERDICT OF SCIENCE. STATEMENTS OF EMINENT MEN.

Dr. Lyon Playfair, M.P., (Professor of Chemistry, Edingburgh), states that, "100 parts of ordinary beer or porter contains $9\frac{1}{4}$ parts of solid matter ; and of this, only six-tenths consist of flesh-forming matter ; in other words, it takes 1,666 parts of ordinary beer or porter to obtain one part of nourishing matter. To drink beer or porter to nourish us, is tantamount to swallowing a sack of chaff for a grain of wheat."

'Alcohol removes the uneasy feeling, and the inability of exertion which the want of sleep occasions ; but the relief is only temporary. Stimulants do not create nerve power ; they merely enable you, as it were, to use up that which is left, and then they leave you more in need of rest than before. It is worthy of notice that opium is much less deleterious to the individual than gin or brandy.'—*Sir B. Brodie.*

'It would not be too much to say that there are, at this moment, half-a-million of homes in the United Kingdom where home happiness is never felt, owing to the cause of tippling alone, where the wives are broken-hearted and the children brought up in misery.'—*Mr. Charles Buxton, M.P.*

'There is scarcely a crime before me that is not, directly or indirectly, caused by strong drink.'—*Judge Coleridge.*

'If you wish to keep the mind clear and the body healthy, abstain from fermented drinks.'—*Sydney Smith.*

'My opinion is, that neither spirit, wine, nor malt liquor is necessary for health ; the healthiest army I ever served with had not a single drop of any of them, exposed to all the hardships of Kaffir warfare at the Cape of Good Hope, in wet and inclement weather, without tents or shelter of any kind.'—*Inspector-General of Hospitals, Sir John Hall, K.C.B.*

'I never suffer ardent spirits in my house, thinking them evil spirits. If the poor could see the white livers, and shattered nervous systems, which I have seen as the consequence of drinking, they would be aware that *spirits and poison* mean the same thing.'—*Sir Astley Cooper.*

'The death from alcoholic poisoning in Great Britain is prodigious; it may be set down at something like one-tenth of the whole death-rate of the country.'—*Dr. Lankester, F.R.S., Coroner for Central Middlesex.*

Professor Gairdner, of Glasgow, proved conclusively that the rise in the death-rate of typhus patients bore a definite proportion to the increase of the doses of alcohol administered, and *vice versa.*

'Beer, wine, spirits, &c., furnish no elements capable of entering into the composition of the blood, muscular fibre, or any part which is the seat of vital principle. 730 gallons of the best Bavarian beer contain exactly as much nourishment as a five-pound loaf or three pounds of beef.'—*Baron Liebig.*

'He had come to the conclusion, from such observations as he had been able to make during many years, that a large proportion of healthy persons, except under special circumstances, were not so well if they took any form or alcohol as they were if they took none; and that conviction was so impressed on his mind that at ordinary dinner parties, especially of young men, it was simply painful to him that custom, mere custom,—these poor youths being really ignorant on the question—made it necessary for them to drink a quantity of beer or wine which, he thought, as far as it affected them at all, was injurious to them. . . . . But as for the old opinion that people in health, and living in ordinary conditions, could not live or work without wine, it was an opinion no careful or thoughtful physician thought of maintaining.'—*Dr. Acland, F.R.S., President of the Medical Council.*

'I had learned purely by experimental observation that, in its action on the living body, this chemical substance, alcohol, deranges the constitution of the blood; unduly excites the heart and respiration; paralyzes the minute blood-vessels; increases and decreases, according to the degree of its application, the functions of the digestive organs, of the liver. and of the kidneys; disturbs the regularity of nervous action; lowers the animal temperature, and lessens the muscular power. Such, independently of any prejudice of party or influence of sentiment, are the unanswerable teachings of the sternest of all evidences, the evidences of experiment; of natural fact revealed to man by experimental testing of natural phenomena.'—*B. W. Richardson, M.A., M.D., F.R.S. : Paper read at the Oxford Medical Conference.*

We cannot prove the safety of moderate drinking by citing the evidence of those who live to old age in spite of it; but we can prove the deadly influence which it has upon the human body by the distinct evidence afforded by the mortuary of any general hospital, which tells us by unmistakable testimony that the person who habitually uses alcohol, as at present supplied, saps the foundation of his health and shortens his life; and that its administration to our children tends to produce a race of individuals who are naturally weak both in mind and body, and who have shorter lives than their fathers. If the use of alcohol as a diet were abolished for all persons under 50 our grandchildren would find the length of life much beyond that which is settled by Dr. Farr as the average to which men now live.—*Dr. Alfred Carpenter.*

'Habitual, or, as it is usually called moderate drinking, is a thing which people should avoid if they wished to have a sound mind in a sound body. That is the reason why I myself touch nothing but water.—*Sir H. Thompson.*

'If there were no such thing as alcohol, half the sin and a great deal of the misery of the world would not be known.'—*Dr. Parkes.*

We cannot regard alcoholic liquors, as contributing to the nutrition of muscular tissues; except in so far as they may contain albuminous matters in addition to the alcohol, which is the case in a slight degree in 'malt liquors.' But these matters would have the same nutritive power, if they were taken in the form of solid food; and the proportion in which they exist in any kind of malt liquor is so small, that they may be fairly disregarded in any discussion on its nutritive value.'—*Dr. W. B. Carpenter, F.R.S.*

PREVENTION IS BETTER THAN CURE.

Dr. BEDDOES says: "There are an infinite number of facts which show that the organization of children is, in general, most apt to suffer from many classes of violent agents. Medical practitioners, much conversant among the poor, find them perpetually stunting the growth and destroying the constitution of their children by their ill-judged kindness in sharing with them those distilled liquors, which they swallow with so much avidity themselves. Among the causes so fatal to the health of the higher classes, the allowance of wine that is served out to children, short as it may appear, deserved to be considered as not the least considerable."

THE LORD BISHOP OF CARLISLE, at an annual meeting in Lambeth Palace Library, April 19th, 1880, speaking of the recommendations of the Committee of the House of Lords on intemperance. said: "I wish to put in a strong expression of our opinion upon the part of that committee, that it would be a very desirable thing to introduce some teaching upon the hygiene of the subject into our elementary schools. This is one of the points upon which, I believe, doctors do not disagree. Anyone who reads the re-published articles from the *Contemporary Review*, will observe that there is certainly a wide divergence of opinion amongst the doctors on this subject; but I think there are two points upon which every medical man in the kingdom is agreed—first, that alcohol is not necessary for children, and in the next place, that it is possible for a man to ruin his health and bring himself to an early grave, by the use of alcohol, without ever having been drunk in his life. Now, I think, that these two truths, if there were no others, if taught to children, would of themselves form the basis of a very important temperance reformation."

Sir William Bovill, said : " Amongst a large class of our population, intemperance in early life is the direct and immediate cause of every kind of immorality, profligacy, and vice, and soon leads on to the commission of crime—including murder, manslaughter, robbery, and violent assault."

Sir A. Carlysle, M.D., says: "The most obnoxious practice is assuredly that of giving children wine and strong drinks at an early period."

Dr. Macnish, writes : " Parents should be careful not to allow their youthful offsprings stimulating liquors of any kind."

Dr. Norman Kerr, of London, writes : " We ought to bring up our children as abstainers, for though we may have been able to drink moderately, one or more of them may be unable to stop at moderation. If they are abstainers till they arrive at years of discretion, they will then be able to judge for themselves, and resolve upon a line of conduct uninfluenced by their previous habits ; but if they are allowed to drink from their earliest youth, a liking for alcoholic beverages may have been acquired which may prevent them from forming an impartial and unbiassed decision."

ENDURANCE OF GREAT BODILY LABOUR WITHOUT BEER.
TESTIMONY OF TRAVELLERS.

The following evidence was given by Mr J. S. Buckingham, in his evidence before the Parliamentary Committee for the Suppression of Intemperance in 1834.—"He once commanded a frigate in the service of the Imaum of Muscat, whose crew consisted of three hundred men, all Arabians, who never tasted any intoxicating liquor ; and they were the most athletic and elegantly-formed men he had ever seen. He has further remarked that when he was at Calcutta he witnessed a trial of strength between a number of men who came down from the Himalayan mountains and the most powerful Europeans who could be selected from the English grenadiers and the vessels in the harbour ; and that in lifting weights, hurling the discus, vaulting, running, and wrestling, each of these Indians was found equal to one and three quarters of the Englishmen ; and yet not one of them had ever tasted any liquor stronger than water." During his extensive travels among the Mahomedan populations of Syria, Mesopotamia, Persia, Egypt, &c., Mr. B. was struck with their generally fine development, and their remarkable amount of muscular vigour, notwithstanding their universal abstinence from alcoholic liquors.

Almost every traveller who has visited Constantinople has been struck with the remarkable muscular power of the men engaged in the laborious out-door employments of that city. Sir W. Fairbairn, an eminent machine-maker at Manchester, remarked, in his Sanitary Report, 1840, p. 252, that " the boatmen or rowers who are perhaps the finest rowers in the world, drink nothing but water ; and they drink profusely during the hot months of summer. The boatmen and water-carriers of Constantinople are decidedly, in my opinion, the finest men in Europe, as regards their physical development; and they are all water-drinkers." And several other observers bear testimony to the extraordinary strength of the porters of Constantinople, who are accustomed to carry loads far heavier than English porters would undertake, even under the stimulus of alcoholic beverages ; yet these Turkish porters never drink anything stronger than coffee.

In the copper mines of Krockmahon, as we are informed by their manager, Captain Petherick, more than one thousand persons are daily employed, of whom eight hundred have taken the Total Abstinence pledge. Since doing so, the value of their productive industry has increased by nearly £5,000 per annum ; and not only are they able to put forth more exertion, but the work is done better, and with less fatigue to themselves. Besides this, they save at least £6,000 every year, which had been previously expended in the purchase of alcoholic liquors.

A similar change, made at the Varteg Iron Works, when under the management of the late Mr. G. Kenrick, was attended with corresponding results ; more work being done in the succeeding six months, than had been ever accomplished within a similar period ; and this being performed with more comfort to the men themselves, and with more satisfaction to their employers.

GREAT FEATS OF ENDURANCE.

SWIMMING.—One of the greatest feats of endurance was that of Captain Webb, who, a short time ago, astonished everybody by swimming across the Channel—from England to France. The time he remained in the water, the distance he swam, together with the continued determination to accomplish his object, were truly surprising, and all this without the aid of ale, wine, or spirits.

WALKING.—Weston, the pedestrian, who walked 450 miles in six days—in one day he walked 96 miles—is another proof of the power of endurance without intoxicating drinks. He did not become tired and footsore, but became drowsy from want of sleep on the sixth day ; but after an ordinary night's rest, when his walk was finished, he got up and went about as though nothing unusual had been undertaken, and has since walked 500 miles in six days.

TRAVELLING.—Who has not heard of Dr. Livingstone, the famous African traveller ? but few are aware, probably, that he braved all, endured all, and did all without using strong drink. This is his testimony :—" I have acted on the principle of total abstinence, from all alcoholic liquors, during more than twenty years. My opinion is that the most severe labours or privations may be undergone without alcoholic stimulants."

SPORTING.—The Messrs J. C. Clegg and W. C. Clegg, of Sheffield, two brothers, studying for the law, who have been total abstainers all their lives, have achieved great results in athletic sports. Mr. J. C. Clegg, in 1868, won six first prizes. The following year he won eight. In 1870, he secured no less than 31 first prizes and two seconds; and in 1872 he won 34 prizes. Altogether Mr. Clegg has won 84 prizes, 74 of which are first. These exploits show that stimulating liquors are not essential to maintain speed, strength, or stamina.

SCULLING.—Mr. Edward Hanlan, the champion sculler, being asked his opinion as to the use of strong drink and tobacco in athletic exercises, furnishes the following reply, dated June 28th, 1879 :—" I have to state that, in my opinion, the best physical performances can only be secured through the absolute abstinence from their use. This is my rule, and I find after three years' constant work at the oar, during which time I have rowed many notable match races, that I am better able to contend in a great race than when I first commenced. In fact, I believe that the use of liquor and tobacco has a most injurious effect upon the system of an athlete, by irritating the vitals, and, consequently, weakening the system."

FIGHTING.—The military career achievements of General Havelock in India, a few years ago, filled all England with admiration ; and yet, during all his severe privations, forced marches, and desperate engagements, he drank neither wine beer nor spirits ; and declared, "that water drinking was the best regimen for the soldier."

EXPLORING.—In the late expedition to explore the Arctic regions, Adam Ayles, and a few others of the party, kept firm to their pledges, and endured all the hardships of that trying enterprise without using any strong drink. Punch notices the heroic conduct, and wishes—

" A health to gallant Adam Ayles,
Who o'er the topers still prevails,
From scurvy safe and Arctic gales,
Through drinking only Adam's ales ! "

Church of England Temperance Chronicle.

BEER *V.* THRIFT AND DOMESTIC ECONOMY.
In a Speech, Town Hall, Wallingford.

T. BLAND GARLAND, Esq., of Burghfield, denied that beer or any other alcoholic drink was necessary. There were excellent substitutes. Beer was a luxury, and nobody should drink it if they could not afford it. And no man could afford it who had made no provision for sickness or old age. Men should be remunerated in some other way than by giving them beer. They had no right to take a man's wages and convert it to poison, and they ought not to educate their boys as drunkards. They tried to emulate the men. People sometimes said there was no harm in a pint a day; but this was five shillings a month. And 3s. 4d. a month, saved by a man 26 years of age, would give him 12s. a week sick pay to the age of 70, and 12s. per week annuity after 70. And a man had no right to his pint a day till he had made some provision for sickness and old age. No young man must look any more to out-door relief. And farmers too often forget that by giving beer in part payment of wages, they first ruin the men and then have to support them. Beer was not really good for them, and it only added to their thirst; in a quarter of an hour a man wanted some more. A substitute should be provided. He then gave his experience. Seven years ago he began to get uneasey about his labourers, and he said to his bailiff " I won't give any more beer." " Then, Sir," he said, " you won't get your hay cut, it will not answer." " Well," I said, " I won't give any more. We will do as much as we can ourselves and the rest we'll leave. The Bailiff was left to make the arrangements. After the harvest, I said, " What about the tea ? " " Why," he said, " the men had done much more work in the time, and had done it better ; I have been a Bailiff for the last 25 years and this is the only year I have been master." He had started a cocoa shop on his farm, and people came again and again. He had also adopted the same plan with his domestic servants, and gave £5 to each man and £3 to each woman in place of beer, and found that he had saved £36 a year by it. He begged those gentlemen who were present to give this a fair trial.

" Let me tell you what happened in my own county. Two years ago, when the Bishop of the Diocese, who always gives one large garden party in the course of the summer, and collects his neighbours from 10, 12 or 15 miles round ; when he gave his garden party he directed his butler that no beer should be given. When the coachmen and servants came they were told there was no beer, only they might have as much tea, coffee and milk as they liked. I asked what was the feeling amongst the men. The reply was " They were all delighted." The butler didn't hear one single grumble or mutter upon the occasion. A great nobleman in the same county did as I believe he has never done before ; the coachmen were told at five o'clock there would be the same five o'clock tea for them that there was for their masters and mistresses. They went and were all just as much pleased as they were on the other occasions."—*Canon Ellison's Speech, Willis's Rooms.*

" For the last two or three years I have gone in for neither beer or beer money. I, on engaging men servants, and my wife, when engaging women servants (and she is thoroughly with me on this matter), put it before them that we do not recognise beer or spirits in any way as a necessity, and there-fore we say—" Your wages will be so much, and if you like to consider so much as wages and so much as beer money you can, but we do not. We give you a lump sum in the form of wages." And though I have never made it a stipulation with any servant that he or she should be a teetotaller, yet both the men and women servants have all very shortly after coming under our roof become total abstainers, and there is not such a thing as beer seen in the ser-vants' hall."—*Charles E. Tritton's Speech, Willis's Rooms.*

HOME-MADE DRINKS FOR THE HARVEST FIELD.

A GOOD HARVEST DRINK.—1 lb. of brown sugar, ½ oz. hops. ½ oz. of ginger (bruised), 1½ gall. of water ; boil the hops and ginger for 25 minutes, add the sugar and boil ten minutes more, then strain and bottle while hot ; it will be ready for drinking when cold, but is better if kept a few days. Dried horehound may be used instead of hops.

GINGER BEER.—1 lb. of lump sugar, 1 oz. of bruised ginger, two lemons sliced ; pour over these ingredients two gallons of boiling water, let it stand till lukewarm, then add one tablespoonful of brewer's barm, or one small teacupful of baker's barm ; let it stand twelve hours, then bottle it. It will be ready for use in twenty-four hours.

INDIAN SYRUP.—1 lb. of lump sugar, 1 oz. of citric acid *powdered*, one lemon cut in slices, one quart of boiling water ; stir all together and bottle when cold. The addition of a little essence of cochineal will impart to it a nice rose colour. One or two tablespoonsful, according to taste, to be put into a tumblerful of water. This will keep more than a fortnight.

MOWERS, HARVESTERS, &c.—The following makes a very refreshing drink, and is almost identical with the above, but much cheaper: 2 lbs. of light brown sugar or more, according to taste, 1 oz. of citric acid, powdered, three gallons of cold water ; mix all together, melt a little common cochineal in hot water and add sufficient to colour it.

ALE OR BEER OUTDONE.—Ingredients : best hops, 1 lb, tapioca, 1 lb., water, ten gallons. Directions: Thoroughly swell the tapioca iu cold water, then put it in a gauze cloth, and boil it and the hops in the water for half an hour. Let all cool together. Next day strain it and put it in bottles or casks, bunging it close to exclude the air. Nothing more is required.—*Irish League Journal.*

By far the best drink is thin oatmeal water, with a little sugar. The proportions are ¼ lb. of oatmeal to two or three quarts of water, according to the heat of the day and your thirst. It should be well boiled, and 1½ ozs. of brown sugar added. If you find it thicker than you like, add another quart of water. Before you drink it, you must shake up the oatmeal well through the liquid. You will find that it not only quenches thirst, but it will give you more strength and endurance than any other drink. In very long harvest days you can take ½ lb. or even ¾ lb. of oatmeal to three quarts of water. You will find this meat and drink also. It must be boiled fresh every morning, or over night will do. You can take it out in kegs or stone jars, and keep it under the shade of the trees, just as you do your beer.

One ounce of coffee and half an ounce of sugar bottled in two quarts ot water and cooled, is a very thirst-quenching drink ; so is cold tea. buf neither of these is so supporting as the oatmeal drink.

It is quite a mistake to suppose that beer or spirits give strength. They do give a spurt to a man, but that quickly goes off, and spurts, in hard, heavy work, too often made, certainly lessen the working powers.

Boiling water poured on a few slices of lemon, with a little sugar, makes a very refreshing drink. Butter milk should be more used as a drink.

The above receipts may be had for circulation, 50 for seven penny stamps, 100 fourteen stamps, post free, from J. Abbey, 44, St. Giles,' Oxford.

* A FARMER'S REASONS FOR DISCONTINUING THE PRACTICE OF GIVING BEER IN THE HARVEST FIELD.

1. Because wages have heretofore been reckoned on the scale of so much per day in money. together with so much beer.

2. Because this is unfair to the labourer. A fair day's work deserves a fair day's wage in money. If I were a labourer I should not like to be obliged to take part in kind ; and as a Christian man I wish to do to others as I would be done by.

3. Because the same rule holds good if I want my men to work overtime. I ought to give them a proportionate increase of pay.

4. Because experience has abundantly shown that the effect of strong drink is to stimulate for the time, not to give real strength for work, and that the hardest work can better be done with the help of nourishing food and cooling drinks than in any other way.

5. Because to attempt to get more work out of them by the means of stimulants than they would otherwise be likely to do, is degrading to the men, and in the end an act of short-sighted policy for myself.

6. Because strong drink in the harvest-field tends, in the case of the men to stupify them, to make them quarrelsome, and to send them at the close of the day to the public house, It is in this way that habits of intemperance are formed

7. Because in the case of women and lads strong drink is utterly out of place. With the latter it leads to habits of beer-drinking in large quantities, and lays the foundation of a life of intemperance and sin. With the former, it leads too often to that female intemperance which has become the disgrace, and is becoming more and more the danger, of our country.

8. Because intemperance, more than any other cause, tends to demoralize the men, and produce a very inferior class of labourer, to the serious loss of employers ; it also leads to crime, poverty, sickness, lunacy, death, and loss of souls.

9. Because the burden of poor and prison rates and of lunatic asylums falls chiefly on the employers of labour, of whom I am one.

10. Because strong drink, being proved to be such a cause of stumbling to others, I am told not " to put a stumbling-block or occasion to fall " before my brother or sister ; and that if I do I am bringing on myself the " woe ' pronounced in Habbakuk ii. 15 ; Matt. xviii. 7.

*AN AGRICULTURAL LABOURER'S REASONS FOR DECLINING TO RECEIVE BEER INSTEAD OF MONEY IN THE HARVEST FIELD.

1. Because " a fair day's work " deserves " a fair day's wage," and the " truck " system, which pays part in money and part in kind, is opposed both to the law of the land and the interests of the labourer.

2. Because this is equally true when working overtime.

3. Because if the attempt is made to get more work out of a man, by stimulating him with strong drink, such an attempt degrades him as a man, and is likely to be followed by injurious consequences to his bodily health and moral condition.

4. Because it is far better to work for my employer's interests as if they were my own, and to labour at all times " with good will, doing service as to the Lord and not to man, knowing that of the Lord we shall receive the recompense of the reward."

5. Because experience has everywhere shown that the man who takes proper food and cooling drinks, can do the work of the harvest field better, and with less fatigue, than the man who is heated and excited by strong drink.

6. Because if this is true of myself, it is doubly true of the women and lads who take part in the work. The craving for strong drink is not natural, but is soon acquired. They will take what is provided for them, and do as the men do to a great extent. Why should I, then, by my example " place a stumbling-block, or occasion to fall," before the members of my own or another's family ?

7. Because strong drink in the harvest-field leads too often to quarrels, bad feeling, ill blood, loss of time and temper, and not unfrequently to actual bloodshed.

8. Because then, and at all times, I wish to have the control over myself, and to be able to do my duty, and to live peaceably with all men.

9. Because intemperance is the curse of my class and my country, degrading us, making us slaves to a vicious habit, producing poverty in our homes, filling our gaols with criminals, our unions with paupers, our asylums with lunatics, and leading its victims to destruction both of body and soul.

10. Because I wish to do my part in taking away this reproach from my native land.

HOME-MADE DRINKS FOR THE HARVEST FIELD.

A GOOD HARVEST DRINK.—1 lb. of brown sugar, ½ oz. hops. ½ oz. of ginger (bruised), 1¼ gall. of water ; boil the hops and ginger for 25 minutes, add the sugar and boil ten minutes more, then strain and bottle while hot ; it will be ready for drinking when cold, but is better if kept a few days. Dried horehound may be used instead of hops.

GINGER BEER.—1 lb. of lump sugar, 1 oz. of bruised ginger, two lemons sliced ; pour over these ingredients two gallons of boiling water, let it stand till lukewarm then add one tablespoonful of brewer's barm, or one small teacupful of baker's barm ; let it stand twelve hours, then bottle it. It will be ready for use in twenty-four hours.

INDIAN SYRUP.—1 lb. of lump sugar, 1 oz. of citric acid *powdered*, one lemon cut in slices. one quart of boiling water ; stir all together and bottle when cold. The addition of a little essence of cochineal will impart to it a nice rose colour. One or two tablespoonsful, according to taste, to be put into a tumblerful of water. This will keep more than a fortnight.

MOWERS, HARVESTERS, &c.—The following makes a very refreshing drink, and is almost identical with the above, but much cheaper: 2 lbs. of light brown sugar or more, according to taste, 1 oz. of citric acid, powdered, three gallons of cold water ; mix all together, melt a little common cochineal in hot water and add sufficient to colour it

ALE OR BEER OUTDONE.—Ingredients : best hops. 1 lb . tapioca, 1 lb., water, ten gallons. Directions : Thoroughly swell the tapioca in cold water, then put it in a gauze cloth, and boil it and the hops in the water for half an hour. Let all cool together. Next day strain it and put it in bottles or casks, bunging it close to exclude the air. Nothing more is required.—*Irish League Journal.*

STOKOS is by far the most refreshing and strengthening drink. It costs 3d. per gallon, and is made as follows :—

Put into a large pan a ¼ lb. of fine oatmeal, from 5 to 7 ozs. of white sugar, half a lemon cut into small pieces. Mix with a little warm water, then pour a gallon of boiling water into it : stir all together thoroughly, and use when cold. The lemon may be omitted, or raspberry vinegar or any other flavoring may be used instead. More oatmeal may be used if preferred.

COKOS is a good nourishing drink made as follows:—8 ozs. of sugar, 6 ozs. of good fine oatmeal, 4 ozs. of cocoa at 10d. per lb . mixed gradually and smoothly into a gallon of boiling water ; take to the field in a stone jar.

One ounce of coffee and half an ounce of sugar bottled in two quarts of water and cooled. is a very thirst-quenching drink ; so is cold tea. but neither of these is so supporting as the oatmeal drink.

It is quite a mistake to suppose that beer or spirits give strength. They do give a spurt to a man, but that quickly goes off, and spurts. in hard, heavy work, too often made. certainly lessen the working powers.

Boiling water poured on a few slices of lemon. with a little sugar, makes a very refreshing drink. Butter milk should be more used as a drink.

The above receipts may be had for circulation. 50 for seven penny stamps, 100 fourteen stamps. post free. from J. Abbey, 44. St. Giles. Oxford.

[OVER.

THIRTY-FIFTH THOUSAND.   THIRD EDITION REVISED   PRICE SIXPENCE.

INTEMPERANCE, ITS BEARING UPON AGRICULTURE;

I.—UPON LANDLORDS.   II.—TENANTS.   III.—LABOURERS.

WITH AN

APPENDIX, containing the Testimony of Landlords, Farmers,
Receipts for Home-made Drinks for the Harvest Field ;

WITH

AN APPEAL TO THE CLERGY,

&c., &c.; BY

JOHN ABBEY.

Extracts from Letters and Opinions of the Press

"Please send me 20 of your excellent Pamphlet. I think it the best production—written or spoken—on the subject I have yet met with."—*A County J.P.*

LADY JOHN MANNERS—Speaking at a Meeting at Knipton, said, "I wish to draw your attention to a pamphlet on "Intemperance': its bearing upon Agriculture!" It contains the testimony of many large employers, who have found by experience that it is better to give wages instead of Beer to their farm labourers, and in harvest time to provide Tea or unstimulating drinks for them. Full directions as to the preparation of these beverages may be found in this Pamphlet." *The Times, April 18, 1882.*

"I will ask you to send me 100 copies of your Pamphlet. I have a very important meeting next week of the clergy, and a great many farmers from over twenty parishes, and I should like to introduce it to their notice. This summer seems to me to be a time when the public mind may be ripe for a great change in the modes of procedure in the harvest field."—*Rural Dean.*

"Mr. Abbey has done good service to Farmers by his earnest appeal in his pamphlet against the use of intoxicating drinks in the hay and harvest field. He is undoubtedly right in his conclusions. It is abundantly proved that more steady and continuous exertion is possible *without* than with intoxicants, and when we consider the moral aspect of the question, the evidence against giving drink is overwhelming. We earnestly commend the pamphlet to the serious consideration of those who still adhere to the vicious principle of supplying beer or cider as part of the wages of their labourers, and we predict from our own experience that when the labourers have made a trial of the new system they will decline to go back to beer."—*The Field, May 20, 1882.*

"Accept my cordial thanks for your most valuable Pamphlet. Nothing could be more serviceable to the great Temperance cause at the present stage of its progress, than to set before employers and employed, and before the clergy and parents a practical review of the facts which experience proves trustworthy."—*Rural Dean.*

"It ought to find its way into every farm house in every parish."—*A Vicar.*

"Kindly forward me 25 copies of your Pamphlet. I think it only requires circulation to become the means of doing much good. Being a land owner, I am anxious that my tenants should perceive the benefits that would accrue to themselves by introducing non-intoxicating drinks into the harvest field."—*A Landlord.*

"Suited in a marked way for distribution in agricultural districts."—*Oxford Times.*

"We should be glad to see it circulated in every homestead in England."—*Oxford University Herald.*

"We hope that our agricultural friends will buy it by the thousands, and thoroughly circulate it."—*British Temperance Advocate.*

"Send me 12 more copies of your Pamphlet; it is doing a good work here. I put a copy into the hands of our largest farmer on Friday last, and on Saturday the copper was going full of tea. The drink is given up."—*A Vicar, June 19.*

"Would do vast good if it could be widely circulated among Farmers, &c."—*London Letter, Alliance News.*

"The pamphlet contains more solid substantial argument in favour of temperance than any other work we have met with."—*Jackson's Oxford Journal.*

For circulation, as follows :—12 for 3s.; 100 for £1.   Apply to JOHN ABBEY, 44. St. Giles', Oxford.